Abraham Metz

The Anatomy and Histology of the Human Eye

Abraham Metz

The Anatomy and Histology of the Human Eye

ISBN/EAN: 9783337395926

Printed in Europe, USA, Canada, Australia, Japan

Cover: Foto ©berggeist007 / pixelio.de

More available books at **www.hansebooks.com**

THE

ANATOMY AND HISTOLOGY

OF

THE HUMAN EYE.

THE

ANATOMY AND HISTOLOGY

OF

THE HUMAN EYE.

BY

A. METZ, M.D.,

PROFESSOR OF OPHTHALMOLOGY IN CHARITY HOSPITAL MEDICAL COLLEGE,
CLEVELAND, OHIO.

Ὑγίεια πρεσβίστα μακάρων.

PHILADELPHIA:

PUBLISHED AT THE

OFFICE OF THE MEDICAL AND SURGICAL REPORTER.

1868.

CAXTON PRESS OF SHERMAN & CO.

PREFACE.

WHEN, a few years ago, I commenced teaching ophthalmology, I seriously felt the want of a text-book on the Anatomy and Histology of the Human Eye. There does not exist, to my knowledge, a treatise on this subject that includes the results of the labors of the more recent histologists to be found in ophthalmological journals and in memoirs on special subjects. It has been my aim to collect this material into a connected form, and in such a manner as to adapt it alike to the requirements of the medical student and of the practising physician.

It affords me pleasure to record my grateful obligations to Dr. S. W. BUTLER for reading the proofs and overseeing the publication of the work, and for the beautiful style in which it is issued, as well as for the obliging disposition he has always manifested.

I am also under many grateful obligations to Mr. HUGO SEBALD, engraver, No. 23 South Third Street, Philadelphia, for his skill and accuracy in executing the engravings.

In conclusion, I venture to hope that my task has been performed in a manner satisfactory to the profession, and that my small book may meet with a friendly reception.

A. METZ.

MASSILLON, OHIO,
June, 1868.

CONTENTS.

ERRATA.

Page 19, top line, read 3 to 4 mm. instead of 3 to 64 mm.

" 48, fifth line from top, read posterior elastic lamina, or membrana Descemeti, instead of *anterior* elastic lamina.

" 56, sixth line from bottom, read membrana Jacobi instead of membrana fucoli.

" 62, fourteenth line from top, read first function instead of first functions.

" 89, tenth line from top, read remains instead of remain.

" 107, eleventh line from top, read portion in place of partition ; also, same page, seventh line from bottom, same correction.

" 108, fourteenth line from top, read Arlt's instead of Alt's.

" 110, ninth line from bottom, read margo instead of marga.

ILLUSTRATIONS.

LIST OF AUTHORS CONSULTED.

PILZ, PROFESSOR JOSEF:
  Lehrbuch der Augenheilkunde. Prag, 1859.

KÖLLIKER, A., Professor of Anatomy and Physiology in the University of Würzburg:
  Microscopical Anatomy. John W. Parker & Son: London, 1860.

RITTER, CARL:
  Die Structur der Retina dargestellt nach Untersuchungen über das Walfischauge. Leipzig, 1864.
  L'Anatomie normale et pathologique du Cristallin, du Corps Vitré et de la Rétine. Written for Wecker's Études Ophthalmologiques. Tome II. Paris, 1866.
  Ueber die Elemente der äusseren Körnerschicht. Archiv f. Ophthal., viii–ii.
  Zur Histologie des Auges. Archiv f. Ophthal., 11–1.

SCHULTZE, MAX:
  Observationes de Retinæ Structura penitiori. Bonnæ, 1859.
  Zur Anatomie und Physiologie der Retina. Bonn, 1866.
  Ueber den gelben Fleck der Retina, seinen Einfluss auf normales Sehen und auf Farbenblindheit. Bonn, 1866.

NUNNELEY, THOMAS, F.R.C.S.E.:
  On the Organs of Vision: their Anatomy and Physiology. London, 1858.

STELLWAG, PROF. DR. KARL VON CARION:
  Lehrbuch der praktischen Augenheilkund. Dritte Auflage. Wien, 1867.

BRUCKE, ERNST:

    Anatomische Beschreibung · des Menschlichen Augapfels.
    Berlin, 1847.

KRAUSE, PROF. W.:

    L'Anatomie et Physiologie de la Conjonctive. Traité Théo-
    rique et pratique des Maladies des Yeux, par L. Wecker.
    Paris, 1867.

LAWRENCE, W., F.R.S.:

    A Treatise on the Diseases of the Eye. American Edition,
    edited by Isaac Hays, M.D.

ARLT, PROF.:

    Ueber den Thränenschlauch. Archiv f. Ophthal., Band i,
    Abth. II.

    Sur les Functions et certaines Dispositions Anatomiques Nou-
    velles du Muscle Orbiculaire des Paupières. Compte-
    rendu de Congrés d'Ophthalmologie. Paris, 1862.

    Ueber den Ringmuskel der Augenlider. Archiv f. Ophthal.,
    Band ix, Ab. 1.

    Zur Anatomie des Auges. Archiv f. Ophthal., Band iii, Ab. ii.

GALEZOWSKI, DR.:

    Etude Ophthalmoscopique sur les Alterations du Nerf Optique
    (chapitre iv). Paris, 1866.

BECKER, DR. F. J. VON:

    Untersuchungen über den Bau der Linse bei dem Menschen
    und den Wirbelthieren. Archiv f. Ophthalmologie, Band
    ix, Ab. II.

IIIS, W.:

    Beiträge zur normalen und pathologischen Histologie der
    Cornea. Bâle, 1856.

MANZ, W. W., Prof. Agrégé a la Faculté de Fribourg:

    Anatomie et Physiologie du Sclérotique et Cornée. Written
    for Wecker's Études Ophthalmologiques.

ARNOLD, J.:

    Die Bindehaut, der Hornhaut, und der Greisenbogen. Hei-
    delberg, 1860.

SCHWEIGGER, DR. C.:

    Ueber den Ganglienzellen und blassen Nerven der Choroidea.
    Archiv f. Ophthal., B. vi, Ab. ii.

IWANOFF, DR. A., aus Moskau:
    Beiträge zur Normalen und Pathologischen Anatomie des
    Auges. Archiv f. Ophthal., B. xii, Ab. i.

MULLER, HEINRICH:
    Anatomische Beiträge zur Ophthalmologie. Archiv f. Oph-
    thal., B. iii, Ab. i.

ZINN, JOHANNIS GOTTFRIED, Professoris Quondam Medici in Uni-
    versitate Gœttingensi:
    Descriptio Anatomica Oculi Humani. Gœttingæ, 1780.

ENGELMANN, TH. WILH.:
    Ueber die Hornhaut des Auges. Leipzig, 1867.

HEIBERG, HJALMAR, DR., in Christianna:
    Zur Anatomie und Physiologie der Zonnula Zinnii. Archiv
    f. Ophthal., B. ii, Ab. iii.

GRAEFE, A. VON:
    Symptomenlehre der Augenmuskellähmungen. Berlin, 1867.

WELLS, DR. JOHN S.:
    Paralytic Affections of the Muscles of the Eye. Ophthal.
    Hospital Reports, vol. ii, p. 44.

FICK, PROF. DR. ADOLF:
    Lehrbuch der Anatomie und Physiologie der Sinnes Organe.

MULLER, HEINRICH:
    Einige Bemerkungen über die Binnen-Muskeln des Auges.
    Archiv f. Ophthal., B. iv, Ab. ii.

LIEBREICH, R.:
    Eine Modification der Schieloperation. Archiv f. Ophthal.,
    B. xii, Ab. ii.

POPE, DR. BOLLING A., Virginia:
    The Nerves and Nerve-Cells of the Choroidea. The Royal Lon-
    don Ophthalmic Hospital Reports, vol. iv, part i, 1863.

HULKE, J. W.:
    Observations on the Growth of the Crystalline Lens. Oph-
    thalmic Hospital Reports, vol. i, p. 182.
    A Contribution on the Amphibian and Reptilian Retinæ.
    Ophthalmic Hospital Reports, vol. iv, part iii.

GRAEFE, A. VON:
> Beiträge der Schiefen Augenmuskeln. Archiv f. Ophthal.,
> B. i, Ab. i.

SAMISCH, DR. TH.:
> Beiträge zur Normalen und Pathologischen Anatomie des
> Auges. Leipzig, 1862.

DONDERS, DR. F. C.:
> Ueber die sichtbaren Erscheinungen der Blutbewegung im
> Auge. Archiv f. Ophthal., B. i, Ab. ii.
> Étude sur les Vaisseaux Visibles à l'extériéur de l'œil. An-
> nales d'Oculistique, tome lii, p. 189.

MOLL, DR. J. A., in Utrecht:
> Bemerkungen über den Bau der Augenlider des Menschen.
> Archiv f. Ophthal., B. iii, Ab. ii.

WITTIG, PROF.:
> Studien über den blinden Fleck. Archiv f. Ophthal., B. ix,
> Ab. iii.

ROSOW, DR. R., in Petersburg:
> Ueber das Körnige Augenpigment. Archiv, ix–iii.

HENKE, W.:
> Die Oefnung und Schliessung der Augenlider und des Thrä-
> nensackes. Archiv f. Ophthal., iv–ii.
> Beleuchtung des neuesten Fortschrittes in der Lehre vom
> Mechanismus der Thränenableitung. Archiv f. Ophthal.,
> B. viii, Ab. i.

KLEINSCHMIDT, DR. A.:
> Ueber die Drüsen der Conjunctiva. Archiv f. Ophthal., B.
> ix, Ab. iii.

LEBER, DR. TH.:
> Anatomische Untersuchungen über die Blutgefässe des
> Menschlichen Auges. Wien, 1865.
> Ueber die Lymphwege der Hornhout. Zehender's Klinische
> Monatsblätter für Augenheilkunde. Januar und Februar
> heft, 1866.

JAEGER, PROF. VON, JR.:

    Ueber die Einstellungen des Dioptrischen Apparates im Menschlichen Auge. Wien, 1861.

COCCIUS, DR. A., Prof. der Med. an der Univ. Leipzig:

    Ueber des Gewebe und die Entzündung des menslichen Glaskörpers.

Fig. 1.

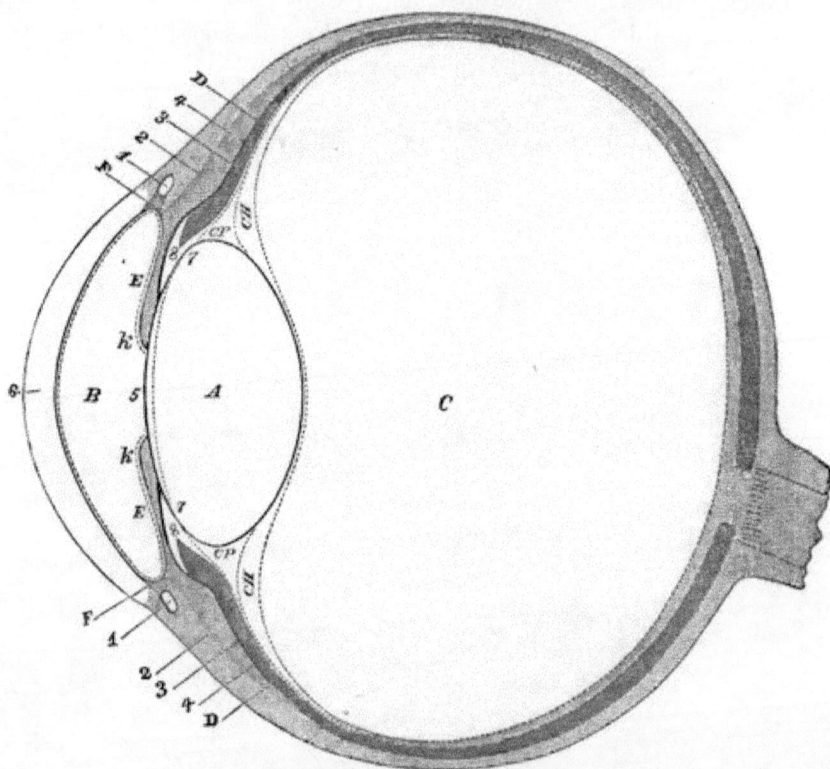

LONGITUDINAL SECTION OF THE HUMAN EYE.

A. Lens. *B.* Aqueous humor. *C.* Vitreous humor. *D.* Apex of the ciliary muscle in the region of the ora serrata. *E.* Iris. *F.* Splitting of the membrana Descemetii to be inserted into the canal of Schlemm in part, and to pass over, in part, to the anterior surface of the iris. 1. Canal of Schlemm. 2. Outer surface of ciliary muscle. 3. Ciliary process. 4. Ciliary muscle near its apex. 5. Pupil. 6. Cornea. 7. Anterior capsule of the lens. 8. Posterior chamber. *CP.* Petit's canal. *CH.* Canal of Hanover. (*From Pilz.*)

MEASUREMENTS.

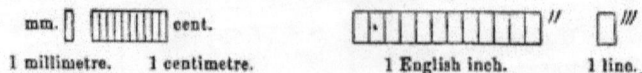

1 millimetre.   1 centimetre.   1 English inch.   1 line.

ANATOMY AND HISTOLOGY

OF

THE HUMAN EYE.

The human eye is an organo-physical apparatus, which has, by means of a system of collective media, the property of casting real images of objects on the retina. The impression of light thus made on the nerve-membrane is conducted by the optic nerve-fibres to the brain, where consciousness is enforced. This wonderful little dioptric organ ranges, anatomically, over a wide field, containing within itself and its appendages all the structures composing the human body. As an optical instrument, it is perfect beyond imitation, having the wonderful property of self-adaptation to long or short distances.

The *visual apparatus* consists of the eyeball and its accessory organs,—the muscles to move the globe, the lachrymal glands to moisten it, and the lids to cover and protect it.

The *eyeball* (*bulbus oculi*) consists of three tunics of distinct structure: a fibrous membrane, the *sclerotica*, with its anterior part or window, the cornea (*tunica cornea*); a vascular and pigment-membrane, the *choroid* (Brücke's *tunica uvea*); a nerve-membrane, the *retina*, on which the luminous impressions are made. Inclosed by these tunics are the *refracting media*, a small bi-convex concentrating lens (*lens crystallina*),

2

inclosed in a transparent membrane, the *lens capsule;* surrounding this are less refracting media, the *aqueous humor,*— a thin, watery fluid, which fills the space between the lens and cornea; and the *vitreous humor (corpus vitreum)* filling the space between the lens and the retina, consisting of a system of membranes confining a humid substance, the *vitrina ocularis.* The enveloping membrane of this is the *hyaloid membrane,* and its continuation forward to the anterior capsule of the lens, like a corrugated or plaited neck, is the *zonula Zinnii.* There are also some transparent membranes, as the outer and inner epithelial coverings of the cornea, and the covering of the choroid and retina, the *membrana limitans.*

The fibrous tunic (tunica fibrosa, sclerotica, cornea).—The outer tunic of the eyeball is a fibrous membrane; its posterior opaque section, which includes four-fifths of the membrane, is the *tunica sclerotica;* its anterior, more convex, transparent section, is called the *tunica cornea.* According to the more recent investigations of Henle, Virchow, and others, the fibrous membrane consists of condensed cellular tissue; the fibrous, or fibrillous structure, seen in investigating the sclerotica, vanishes after immersing the eye for some time in hot water, until the membrane feels hard. The sclerotica is opaque, white, hard, unyielding, and poor in vessels. It is the skeleton of the eye; it serves for the attachment of muscles, sustains the form of the eye, serves for the transmission of vessels and nerves to the parts within, envelops the fluids and protects them. Anteriorly the sclerotica terminates in the cornea. This change is never abrupt, but the line of demarcation is cloudy and semi-transparent. The union seems to take place by bevelled surfaces, the sclerotica overlapping the cornea externally, so that the anterior chamber extends further back than does the cornea on the outer surface. On its posterior surface, the sclerotica is perforated by the optic nerve, the sheath of the latter being continuous with the sclerotica, and similar in structure with it, although much thinner. The optic nerve entrance is a little to the inner side

of the eye axis; according to Hasner, 3 to 64 millimetres. The
optic nerve sheath consists of two layers, an inner fibrous
layer, an outer fibrous layer, with a layer of cellular tissue
between them. Its structure will be considered hereafter, and
is referred to here to aid in the description of the optic nerve
entrance, which is a point possessing considerable pathological importance in connection with glaucomatous conditions.
Donders says, that of the two layers of the optic nerve sheath,
the outer, with its vessels and nerves, penetrates two-thirds of
the thickness of the sclerotica, whilst the inner layer immediately proceeds behind the choroid, with which some of its
fibres unite, and turns outward, and is expanded on the inner
surface of the sclerotica. From it, however, a number of elastic elements proceed, and pass between the individual optic
nerve bundles, to form the so-called *lamina cribrosa*, which is
connected with but a small portion of the choroid. At this
point the optic nerve is only and exclusively surrounded by
the firm, cellular texture of the *sclerotica*. The *arteria* and *vena*
centralis retinæ also pass through the *lamina cribrosa*, causing
the largest opening in the cribriform lamina, which is sometimes called *porus opticus*. Around the optic nerve entrance
the sclerotica is also perforated by the long and short ciliary
arteries, some venous branches, and by the ciliary nerves. In
the region of the *equator bulbi* it is perforated by the *vasa*
vorticosa of the *chorioidea*, and around the corneal border the
arteria and *venæ ciliares anticæ* perforate it.

The anterior portion of the sclerotica is covered by the tendons of the recti muscles, but their expansions do not meet so as
to form a continuous membrane, but are only, in a manner, connected by the *tunica vaginalis bulbi*, or *Bonnet's capsule*, which
begins at the foramen opticum, envelops the globe loosely,
to a point anterior to the equator, where the recti muscles
perforate it, and it is blended with their sheaths and the expansion of the tendons, and is firmly attached to the sclerotica
as far forward as the border of the cornea. This latter portion
of the membrane is called *Tenon's membrane*. This capsule

will be more fully described in another place. The sclerotica is firmly connected with the choroid in the region of the ciliary muscle, and at the optic nerve entrance. In the rest of its extent it is but slightly connected by means of delicate cellular tissue, the *lamina fusca*, which contains some nuclei. On separating the choroid and sclerotica, some of the pigment contained in the *lamina fusca* is torn loose, and adheres to the sclerotica. Pilz says that the pigment contained within the *lamina fusca* is sometimes increased at the immediate border of the optic nerve entrance, and causes the dark ring sometimes seen in using the ophthalmoscope. The sclerotica is thickest posteriorly, being a little more than $\frac{1}{2}'''$. It is thinnest at the equator, being only $\frac{1}{4}'''$ to $\frac{1}{5}'''$, and near the corneal border it is $\frac{2}{5}'''$ in thickness. In structure, the sclerotica is composed altogether of bundles of cellular tissue, which cross each other in longitudinal and transverse layers, the former predominating posteriorly, and the latter anteriorly. On the other hand, the cornea is complicated in structure. The *cornea proper* is an immediate continuation of the sclerotica, also consisting of condensed cellular tissue, but not arranged in longitudinal and transverse layers, as in the sclerotica, but the bundles run in different directions, and its fibres are so firmly interlaced that it has the appearance of compressed sponge. In boiling, corneal substance yields chondrin, whilst the sclerotica yields glue. The cornea is transparent, yet not clear as crystal, whilst the sclerotica is quite opaque. The cornea is colorless, elastic, and resistant, whilst the sclerotica is white, less elastic, but quite firm. The cornea may somewhat easily be torn into lamellæ, in consequence of its bundles of fibres being parallel, whilst the structure of the sclerotica does not permit it to be torn into layers. As Stellwag truly says, the lamellar structure of the cornea must be asserted, whatever the microscopist may say, as is seen in the regular splitting of the corneal layers in onyx, and in interlamellar extravasation of blood. It has been noticed above, that the cornea and sclerotica are united by bevelled surfaces, the latter overlapping the former externally.

This union is so firm that even macerating cannot separate them. In fact, the cornea proper, or lamellated cornea, and the sclerotica, are one continuous membrane. We owe to William Bowman, of London, a clear and full description of the cornea. It is an elaborate structure, and consists of five coats or layers, that can clearly be distinguished, which are, from before backward, the *conjunctival layer of epithelium*, the *anterior elastic lamina*, the *cornea proper*, the *posterior elastic lamina*, or *membrana Descemeti*, with its epithelial covering. The cornea *proper*, or *lamellated* cornea, constitutes the main thickness and strength of the cornea. It is a modification of the white fibrous tissue, immediately continuous with that of the sclerotica. In the sclerotica the fibres run somewhat regularly, in longitudinal and transverse directions, whilst in the cornea they flatten out into a membranous form, and follow mostly the main curvatures of the corneal surface, and constitute a series of more than sixty lamellæ, intimately united to one another by very numerous processes of a similar structure, passing from one into the other, and making it impossible to trace any one

Fig. 2.

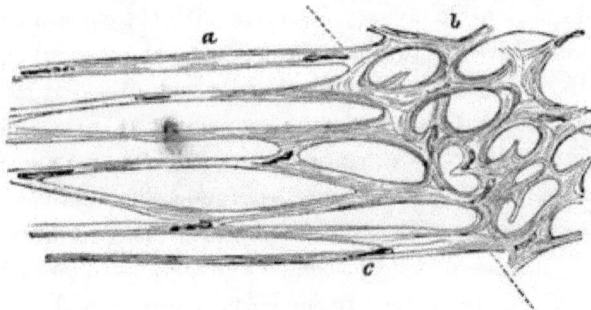

Vertical section of the Sclerotic and Cornea, showing the continuity of their tissue between the dotted lines. *a*. Cornea. *b*. Sclerotic. In the cornea the tubular spaces are seen cut through, and in the sclerotic the irregular areolæ. Cell-nuclei, as at *c*, are seen scattered throughout, rendered more plain by acetic acid. Magnified 320 diameters. (*From Todd and Bowman.*)

lamella over even a small portion of the cornea (Bowman). In the sclerotica there is a network of the finest fibres, of an elastic elementary character, leaving at the places where the original areolæ were located, spindle-shaped and stellate spaces, forming channels, carrying a thin, nutrient plasma. In corneal substance there is between the bundles of connective tissue, a less developed and less branching elastic tissue, in the form of freely anastomosing spindle-shaped and stellate nucleated cells. These are the "cornea corpuscles." According to Langhans, these corneal bodies are much more numerous near the corneal border.

The minute structure of the cornea has been the object of laborious and patient investigation, and yet the published results of the most recent histologists prove that the work is far from being perfected. Many points connected with its minute structure are now the subject of ardent discussions. Engelmann (*Ueber die Hornhaut des Auges*, Leipzig, 1867) says, that the cornea proper is composed of the finest fibrillæ, which lie close to each other, and lying between these are numerous cells and nerves. In the frog these fibrils have a thickness of 0.0001 mm., and each one is separated from its neighbor by an immeasurably small space filled with a fluid. These fibrillæ are united into larger lamellæ about 0.004 mm. in thickness, which are placed into 15 to 20 layers concentric with the corneal surface. The fibres of each layer run parallel with the corneal surface, and with each other. The fibres cross each other, at an angle of about 90°, in two contiguous layers resting on each other. In some places the fibres run from one lamella into another.

Between two contiguous layers are found, distributed at equal distances apart, a large number of cells. These corneal cells consist of masses of *protoplasma*, polygonal in form and without nuclei. They are vertical to the corneal surface and flattened. In the centre of each mass is found a vesicular nucleus with a nucleolus. These masses that surround the nuclei measure 0.02 mm., and have projecting from their corners from six to twenty processes, which run in various directions throughout the corneal substance. The majority do not project beyond two contiguous lamellæ; some, however, pass through the lay-

ers at sharp angles. Some of these processes terminate free in
minute points; others are connected with neighboring cells.
Each cell then is connected with other cells in the same layer,
and also with the cells of the layers above and below, so that
the whole corneal substance is connected by a penetrating network of this protein material. Neither the cells nor processes
have membranes, but lie unenveloped within the inter-fibrillar
spaces, which they completely fill.

Besides the stellate bodies existing in the true corneal substance, there are found, normally, a number of smaller cells,
without a membranous envelope, which constantly change their
form and locality, called the "*wandering cells.*" They are found
in all the layers of the cornea proper, and do not move about
in bounded channels, but in the interspaces filled with fluid
between the corneal fibres. They seem to push aside the *fibrillæ*
that come in their way. These "wandering cells" are found
floating through all parts of the *substantia propria*, which is certain evidence that the interspaces between the fibrillæ are filled
with a liquid substance. This understanding of the histology
of the cornea excludes a system of closed nutritive channels,
as taught by Von Recklinghausen, Bowman, Sämisch, Leber,
etc., which Engelmann claims to be artificial dilatations from
the injections used. This also excludes lymphatic vessels from
the cornea. He bases his claims of superior success in this investigation on the fact that he made his observations on entirely fresh corneæ without any hardening preparation or injections. The corneæ were moistened only in the aqueous humor,
and examined a few minutes after separation from the living
frog.

The "wandering cells" cannot pass through the anterior
elastic lamella, which is more firm in texture than the *substantia propria*. There are found "wandering cells" in the anterior corneal epithelium, which exist in great abundance.
They are, however, of quite another kind, unlike those existing in the cornea proper, being much smaller than the latter,
and possess generally from two to three nuclei. Their length

is about 0.01 mm., and their protein substance is granular. In moving about, they pass through between the epithelial cells.

These masses of plastic material, containing cells with nuclei, and having anastomosing processes, have been the subject of most ardent discussions; and whilst Engelmann's method of investigation on fresh, *living* corneæ seems to possess important advantages over the methods of hardening, coagulating, and injecting, as practised by other histologists, the time has not arrived for a decision on the character of these "cornea-corpuscles" of Virchow. It is claimed (Ilis, Kölliker) that the nutrient canals of Bowman are artificial productions brought about by his injections.

Hasner, Virchow, Strube, Luschka, and others, believe these nutrient channels to be lymphatic vessels. Recklinghausen and Leber, in quite recent publications, believe them to be *lymph channels*, having a direct communication with the *lymph vessels of the conjunctiva*. Leber made use of turpentine and *dragon's blood* for injecting material, and made an incision near the centre of the cornea. After some minutes the coloring matter injected was seen following certain channels in the conjunctiva, which he thinks were the lymphatic vessels. Leber also claims that these channels possess a distinct enveloping membrane of their own. This much seems to be established concerning these cornea-corpuscles,—they are connected with the nutritive process of the cornea.

The anterior elastic lamina (stratum Bowmanni) is found between the proper cornea and the conjunctival epithelium. At the present day some doubt its independent existence. At any rate it seems more rudimentary than the posterior elastic lamina. Arnold considers it similar to the basement-membrane of mucous tissues. It is coextensive with the cornea, and Manz says that it is lost at the border thereof (certainly at the superior and inferior borders), in the manner of the posterior elastic lamina, to become blended with the cellular tissue of the conjunctiva. According to Bowman it is a vitreous membrane, perfectly clear, with a tendency to curl inward, and it materially aids the cornea in retaining its exact curvature. From its posterior surface a number of cords proceed into the proper cornea, and into the sclerotica, which bind it down firmly to those tissues. According to Arnold it has a thickness of 0.0045 mm. On its anterior surface it is covered by the *conjunctival epithelial layer of the cornea.* This consists of three layers: the first, a hyaline lamella of fused cells; beneath this, a layer of cylindrical nucleated cells placed upright and closely arranged; and lastly, the layer in immediate contact with the anterior elastic lamina, which consists of a layer of round cells, with large nuclei, imbedded lightly in a viscid fluid.

FIG. 3.

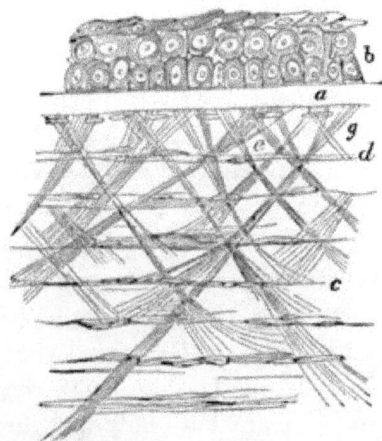

Vertical section of the Human Cornea near the surface. *a.* Anterior elastic lamina. *b.* Conjunctival epithelium. *c.* Lamellated tissue. *d.* Intervals between the lamellæ showing the position of the corneal tubes collapsed. *e.* One of the nuclei of the lamellated tissue. *g.* Fibrous cordage sent down from the anterior elastic lamina. Magnified 300 diameters. (*From Bowman.*)

The thickness of this layer is 0.01–0.05′′′. It can easily be scraped off from the layer beneath it, and is readily regenerated.

The posterior elastic lamina (membrana Descemeti, seu Demoursi) is a perfectly transparent structureless membrane, which extends as far as the cornea, then divides into numerous cords or fibres, which partly terminate at the canal of Schlemm, and partly on the anterior surface of the iris, in a manner to be hereafter described in connection with the vascular membrane. It is a vitreous membrane, very hard, but easily torn. It has a great tendency to curl towards its concavity, and assists in retaining the proper curvature of the cornea, essential to perfect vision. It is ordinarily called *membrana Descemetii*. It has a posterior epithelial lamella, consisting of a single layer of round cells with large nuclei.

The cornea in its centre has a thickness of about 0.436'''. Helmholtz says it is of uniform thickness until near the periphery, where it is increased.

Under intraocular pressure the cornea is flattened and its radial curvature is increased. The cornea may be compared to a section of a smaller globe attached to a larger, and the greater the pressure against the walls of the eye from within, the nearer does the bulb approximate a perfect globe. The posterior or inner wall of the cornea is round, and has a diameter, in all directions, of 5''', whilst its external or outer wall measures less in consequence of the manner of its union with the sclerotica, its vertical diameter being 4½''', and its transverse diameter nearly 5'''. In penetrating the anterior chamber close to the periphery of the iris, as in the operation of iridectomy, it is necessary to make the incision 1''' behind the corneal border above, ¾''' below, and ½''' at the sides. The cornea is hard, resisting, difficult to penetrate, but easily split into layers, a fact realized by all young operators.

The fibrous tunic is poor in vessels, being nourished by a peculiar system of nutrient canals. The arteries from which this nutrient plasma is drained for the sclerotica, are the *vasa ciliaria anticæ*, which proceed to the iris and ciliary body; posteriorly the *arteriis ciliare posticis brevibus*, which also mainly supply the choroid.

Bloodvessels of the Cornea.—It has been generally taught by anatomists that the cornea is nourished from the palpebral and lachrymal arteries, which form the superficial layer of vessels, and project forwards about 1''', where they form loops. According to a recent publication by Leber (*Anatomische untersuch-ungen über die Blutgefässe des Menschlichen Auges*, Wien, 1865), this seems to be a mistake. By careful and successful injections he has demonstrated that it is the anterior ciliary arteries which form the loops at the margin of the cornea, and that the palpebral and lachrymal arteries do not reach as far forward as the *annulus conjunctivæ*. The small branches springing from the fine loops of the anterior ciliary arteries run back some distance to meet and anastomose with the peripheral arteries of the conjunctivæ, so that a belt of vascular network is formed around the corneal border from which the branches of the latter (peripheral) are excluded. A more full description of the vessels supplying the cornea with nutritive plasma is given in connection with the vascular system of the eye.

As regards the *serous vessels* of the cornea, described by Arnold, Hasner, and others, to say the least, their existence has not been satisfactorily demonstrated. Leber says that, in the perfectly healthy eye, he has never been able to discover these supposed serous vessels, and he does not believe in their existence. He further adds, that the matter does not deserve the importance ordinarily attached to it, inasmuch as it is an established fact, that the nutrition of the tissues does not take place directly through the bloodvessels, but through lymph-channels permeating through them; and hence it cannot be of much physiological importance, whether at the corneal border there are or are not such fine vessels capable of circulating serum only.

Nerves.—It has not been positively demonstrated that the sclerotica possesses nerves. Bochdalek believed that he had traced nerves into its tissue; Luschka, Kölliker, and others, could not succeed in finding them. The cornea, however, is plentifully supplied with nerves. Kölliker says 20 to 30 twigs from the ciliary nerves run along the proper cornea; the fine branches, however, are directed forward, and in the corneal centre a network is formed by free anastomoses, immediately beneath the anterior epithelium.

Schlemm, in 1830, first published the presence of nerves in
the cornea. Since then, the observations and opinions on those
nerves have been extensive and varied. Pappenheim, Luschka,
Valentin, Bochdalek, Engel, Beck, Purkinje, etc., have published the results of their investigations, and, quite recently,
His, Sämisch, Kühne, Cohnheim, and Engelmann, have added
much to our knowledge on this subject. According to His and
Sämisch (with whom Kölliker, Coccius and Arnold agree), the
nerves of the cornea are derived from the ciliary nerves and
from the nerves of the *conjunctivæ oculi*, and form numerous
threads, which, at first, have a double contour, which, however,
change to a single contour, after the first division. These nerve-
fibres are very pale, and present, from the origin of their course,
numerous nuclei, located within an extremely delicate *neurilemma*. These nuclei become more rare, the fibres become more
pale, and divide dichotomously in such a manner that the two
branches of the bifurcation run in nearly opposite directions.
The secondary nervous fibres anastomose, and form a network
located quite superficially in the cornea, where His and others
saw the termination of the nerves composing it. Sämisch observed the anastomoses of the secondary fibres, as well as communications between the fibres of the first order. He thinks
that the terminal plexus is not constituted by the nucleated
fibres of His, but by a network of still finer fibres. Manz,
Krause and Kühne have asserted that the plexus is not the
termination of the nerve-fibres, but that extremely fine, pale
fibres end free, after a certain course. Kühne thinks the free
extremities belong to the order of motor nerves, and communicate with the emanations of the cornea corpuscles. He says:
"Les cylindres axes nus, qui sortent enfin de ses divisions multi-
ples, deviennent légèrement granuleux et se combinent continu-
ellement aux filaments du protoplasme de corpuscules de la
cornée. Ainsi il est probable, qu'il n'y a pas un seul corpuscule
(cellule) de la cornée, qui ne soit en combinaison directe ou indi-
recte avec les elements nerveux. Quand au rôle des ces nerfs,
nous avons constaté, qu'ils sont une espèce de nerfs moteurs."

(*Gazette hebdomadaire*, tome ix, No. 15.) The triangular expansions at the nodal points of the nervous plexus, regarded by

FIG. 4.

The coarser ramification of the corneal nerves of a new-born child, outer half of the cornea of the right eye. As far as the stems are dark-colored, they contain nerves of double contour. The net-like connection of the fibres is not seen in this degree of enlargement. Magnified 14 diameters. Prepared with acetic acid. (*From Samisch.*)

Coccius as *true* ganglia, Sämisch considers merely expansions of the primitive fibres.

Engelmann's observations differ from Kühne. He says that from the border of the sclerotica, the distance of two or three laminæ outward from the membrana Descemeti, are found a large number of pale nerves with dark borders, which pass into the cornea. A portion of these are to end within the true cornea, and another portion penetrate the *elastica anterior*, to end in the anterior epithelium. The dark-bordered fibres, visibly medullated, are united into larger branches, pass the border of the sclerotica, to penetrate the cornea at six or eight places. Each of these stems contains from five to fifteen dark-bordered nerve-fibres. At the distance of about 0.3 to 0.5 mm. from the border of the sclerotica, the medullary part of the nerve vanishes, and in their further progress they are completely transparent and clear non-medullated fibres. In the second or third lamina from the *elastica posterior*, these nerves ramify in all directions, so that any one of the larger branches are connected with all the other branches. As long as the fibres possess the dark-colored medullary layer, they divide only exceptionally; as soon as they become non-medullated clear fibres, they very rapidly divide, and become finer. The fibres in the network, in the true cornea, form no true anastomoses, but connections by contiguity only are found. Each dark-bordered fibre is enveloped by a delicate sheath, containing nuclei. This sheath, getting thinner all the time, is also continued on the pale fibres. The nuclei are numerous near the corneal border, but diminish toward the centre of the cornea; they are found only in the crossings of the plexus.

The corneal substance is not alone supplied by the dark-bordered nerves; there are also quite a number of very fine pale fibres, which enter the corneal border from the sclerotica. At many points at their entrance into the cornea they are united, or, at least, in contact. In number they often amount to sixty. They generally enter the cornea in the second lamella (counting from within), and mostly remain in this lamina. They are

distinguished by their extreme fineness and want of nuclei or
visible sheaths. They seem to end mostly in the posterior
lamina of the cornea. All that Engelmann can positively say
concerning the termination of these fine fibres, as well as of the
dark-bordered fibres, is that they are seen as extremely fine
pale fibres, until they can no longer be seen in consequence of
their great fineness.

A large portion of the dark-bordered nerves that enter the
cornea from the border of the sclerotica, proceed to the anterior
epithelium of the cornea. From the coarse network formed
by the ramification of the nerves in the proper cornea, as above
described, are given off numerous nervous branches which run
forward and penetrate the *elastica anterior*. They are given
off mostly from the crossings of the plexus, sometimes at a
true angle, and seldom less than at an angle of 60°. In number they vary from forty to sixty for each cornea. The larger
of these branches are either bundles of the finest pale nerves,
or undivided pale thick *axis-cylinders*, which latter are found
mostly near the corneal border. When the nerve-fibres have
reached the *elastica anterior*, they perforate it in a vertical
manner. The perforations made by the nerve-fibres through
the anterior elastic lamina, Engelmann names *nerve-pores*. No
twigs are distributed in the elastic lamina. All the fibres that
have penetrated the *elastica anterior* consist wholly of the continuation of axis-cylinders. They ramify without nuclei or
sheaths, beneath and between the cylindrical cells, the layer
immediately in front of Bowman's membrane. They ramify
freely between and beneath the cylindrical cells, so that a somewhat close nervous network is formed. They do not appear to
reach the anterior layer of flattened cells, but terminate free
between the cells of the middle or posterior layers of the
epithelium, without end organs or a terminal network. As
already stated, the fibres terminate invisibly fine, and free
among the cells. It seems (according to Engelmann) that the
latest published results of Kölliker's (*Ueber die Nervenendigun-
gen der Hornhaut. Aus der Würzburger naturwissensch.*, Bd. vi,

1866) investigations of this subject coincide mainly with the above views of Engelmann.

Kölliker also says that there are both an intra-corneal as well as an epithelial expansion of nerve-fibres. He, as also Engelmann, could not find any connection between the intra-corneal nerves and the cornea-corpuscles.

The sclerotica may be compared to an ellipsoid, with its anterior segment cut off. When the anterior, open part is considered a plane, then its antero-posterior axis measures about $9\frac{1}{2}'''$, its vertical diameter at the equator is $10'''$, and the horizontal $10\frac{1}{4}'''$.

The curvature of the outer corneal surface is considered as the segment of a rotational ellipsoid, with a radius of $3.456'''$, whilst the curvature of its inner wall has been considered as the section of a revolutional paraboloid with the half parameter of $2'''$ (Pilz, Volkmann, Kohlrausch, Stellwag).

The cornea is the most important part concerned in the refraction of light. Its anterior surface may be considered the section of a globe, with a radius of about $3\frac{1}{2}'''$ (Hasner). Its refracting index is, according to Krause, 1.342, and is nearly equal to that of the aqueous humor. It seems that its posterior surface, which is frequently irregular, has but little influence on the passage of the rays of light. Stellwag says the cornea casts a focus $5'''$ behind the retina. Hasner gives its posterior focal distance as 30.61 mm. (a little less than $15'''$); and its anterior focal distance 22.81 millimetres (a little less than $11'''$). The cornea is firm, and does not readily change its curvature. Its power in directing the rays of light toward the axis of the eye is great, acting as the object-glass of the eye, and the lens as a collecting-glass.

The Choroid (Choroidea, Tunica Uvea).—*The vascular tunic* of the eye extends from the entrance of the optic nerve to the pupil, thus lining all of the interior of the eye, with the exception of the optic nerve entrance and the anterior chamber. It consists of three divisions. The *first* and largest division extends from the optic nerve entrance to a point a little in front of the equator, and 2, 5''' to 3''' posterior to the anterior border of the sclerotica, in the vicinity of the *ora serrata retinæ*, and is called the *choroidea*.

The *second* division begins at the *ora serrata retinæ*, where the vascular membrane becomes thicker, and extends forward 2, 5''' to 3''' to the junction of the sclerotica and cornea, called the *corpus ciliare*.

The *third* division begins at the anterior border of the *corpus ciliare*, and proceeds vertically toward the axis of the eye, as far as the pupillary margin, and is called the *iris*.

The bulk of the *choroidea* consists of a close, vascular network, imbedded in a stroma of cellular tissue.

The arterial supply of the vascular membrane (*choroidea*, ciliary body and iris) is derived from the ciliary arteries, the *posterior ciliary arteries* being direct branches of the ophthalmic artery, and the *anterior ciliary arteries*, which are branches from the arteries supplying the straight muscles of the eye. We are indebted to Leber for a clear description of the bloodvessels of the vascular membrane, and we shall follow him in our descriptions of the vessels within the eye.

The posterior ciliary arteries must be further divided into the *short* posterior *ciliary arteries*, which are distributed only within the choroid itself, and the *long* posterior *ciliary arteries*, which proceed directly to the ciliary muscle, running in their course between the choroid and sclera. In the ciliary muscle the long posterior ciliary arteries form a connection with the anterior ciliary arteries, to supply with blood the *corpus ciliare, the iris, and the anterior portion of the choroidea*. The vascular membrane, then, has an anterior and a posterior system of bloodvessels, which are not independent of each other, but form free connections. The anterior region is supplied by the long posterior and the anterior ciliary arteries, and the posterior region is supplied by the short posterior ciliary arteries.

The short posterior ciliary arteries proceed from the ophthalmic artery in 3 or 4 small branches, which pass to the posterior part of the sclerotica, around the optic nerve, to divide into numerous

branches, to be distributed to the choroid, the posterior part of
the sclerotica, and the optic nerve at its entrance into the sclerotica.   The branches destined for the choroid are about 20 in
number, which perforate the sclerotica in a direct manner
around the optic nerve. After having perforated the sclerotica,
the short ciliary arteries begin to divide, mostly at acute angles,
into smaller branches. They run for some distance in a tortuous manner, in the outer layer of the choroid, but soon pass inside of the thick layer of the veins into the deeper layers of
the membrane. During their course they constantly give off
branches to the inner layer of the choroid,—the capillary network, where their terminal branches finally disappear. The
larger the branches are, the further forward they proceed for
distribution. The small branches that penetrate the eye close
to the optic nerve, supply only the posterior part of the choroid;
whilst the larger branches pass forward with its twigs, as far
as the capillary network (*membrana chorio-capillaris*) which extends to the *ora serrata*. Only a few of the branches of the
short ciliary arteries pass beyond this limit, to anastomose with
the branches of the anterior and long ciliary arteries, to form a
connection with the anterior and posterior systems of vessels.

For years the text-books have been following Brücke in the
manner of distribution of the bloodvessels of the *choroidea*,
who describes an outer layer of arteries which do not terminate in capillaries, but after having, through division, reached
a certain degree of fineness, they curve around, to end immediately in the branches of the *venæ vorticosæ*, to form an anomaly in anatomy. Then Brücke has an, inner layer of arteries,
which form the close capillary choroideal network. Also, an
anterior layer of posterior short ciliary arteries, which extends
from the capillary network to the roots of the ciliary processes.
Leber denies the existence of this arrangement of distribution
of the posterior short ciliary arteries, and his patient and successful investigations in that direction seem to deserve cre-
dence. After repeated injections of the arteries with glycerine
and sulphate of baryta, and the veins with glycerine and Berlin

blue, he uniformly found that the injected material of the arteries could only reach the veins *through capillaries.*

Membrana Chorio-capillaris.—The capillary network forms the inner layer of the *choroidea,* and extends from the entrance of the optic nerve to the beginning of the non-fimbriated part of the ciliary processes. Its meshes, or interspaces, are always radiary, being always located at the terminal extremity of an artery or a vein,—*i. e.,* where an artery terminates by division into capillaries, or a vein begins by the union of capillaries, giving rise in this manner to the beautiful star-shaped figures seen on the choroid. The meshes are finest in the posterior division of the choroid, and become larger further forward. In the region of the *ora serrata retinæ* the capillary network ceases, and only a few capillaries are found between the straight veins of the smooth portion of the ciliary processes.

Veins of the Choroid.—The blood from the vascular tunic is carried off mainly through the *venæ vorticosæ.* A small portion escapes through the *anterior ciliary veins.* The posterior *short ciliary veins* consist of a few fine venous trunks, which carry off blood only from the sclerotica, and *none from the choroid,* as has been taught by Brücke and others. The *long posterior ciliary veins,* corresponding with the *long posterior ciliary* arteries, as described in text-books, *do not exist.* The *anterior ciliary* veins will be described hereafter.

The *venæ vorticosæ* consist of four trunks, which proceed to the equator of the eye, and, before entering the sclerotica, sometimes divide into two branches, so that sometimes six are counted. They pass out of the sclerotica very obliquely, making channels of 1½ mm. to 5½ mm. in length. In the choroid, the veins form the well-known whorls, as the veins approach it from all directions in a radiary manner. The larger, 4 to 6, branches, form complete *vortices;* the smaller form only imperfect ones. The trunk of a *vortex* receives its larger branches from all sides ; those coming posteriorly and from the sides are from the *choroidea;* those from the front are from the ciliary body and the iris. It is not difficult to conceive in what manner

these vortices are formed. The curved manner in which the veins
have, necessarily to reach the vortex, accounts for the radiary
arrangement of those veins.   Where the distance between the

FIG. 5.

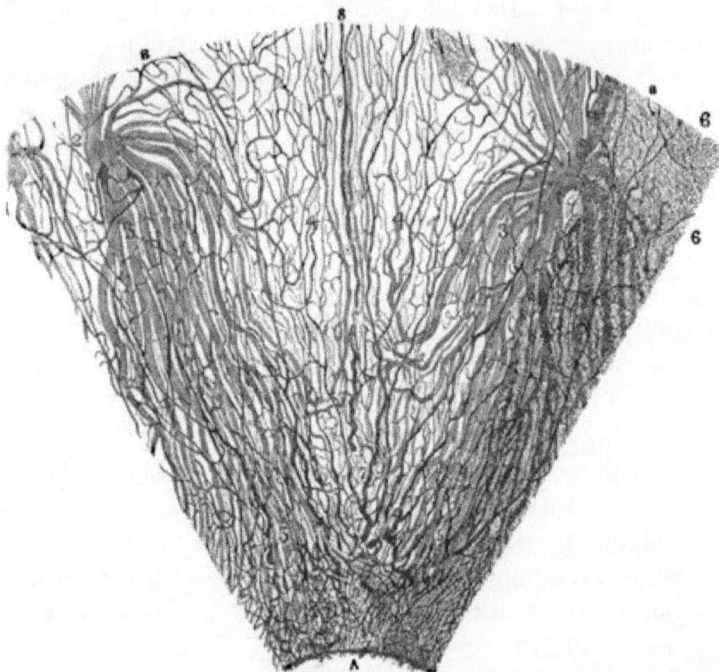

FIG. 5. A, Border of the optic nerve. B, B. Equatorial region of the choroid. 1. Perfect vortices. 2. Imperfect vortices. 3, 3. Anterior venous connecting bows or curves.
4, 4. Parallel veins which enter the first connecting curving veins. 5, 5. Posterior connecting venous bows or curves. 6. Capillaries of the choroid. 7, 7. Large branches of
the posterior ciliary arteries. 8. Long posterior ciliary artery. (*From Leber.*)

vortices is great, the veins midway between them do not bend
over, but proceed in a straight manner, as seen at 4, 4, Fig. 5,
until they reach a curving vein, with which they unite.

The arteries and veins in the choroid have, by their close
contiguity, a mutual pressure on each other. By turgescence
of the arteries, the veins are compressed, and by engorgement
of the veins, pressure is made on the arteries.

The *choroidea* is soft, yielding, and easily torn. In its posterior part it has a thickness of $\frac{1}{15}$ to $\frac{1}{20}'''$; in the region of the equator, $\frac{1}{30}'''$. Anteriorly it is a little thicker again.

The woof (*texture, stroma*) of the choroid is a tissue intermediate between cellular and elastic tissues. In some localities the one, and in other localities the other, predominates. The external part of the stroma of the choroid, in which the vessels run, consists of nucleated, spindle-shaped and stellate cells, which are quite irregular, colorless, or brown, with delicate processes of variable length, which interlace freely, so as to form a loose membrane, somewhat resembling the fibro-elastic membranes. In the inner layer of the choroid the connective tissue is less pigmented, and passes over into a homogeneous nucleated tissue, quite similar to the elastic lamella of the inner

Fig. 6.

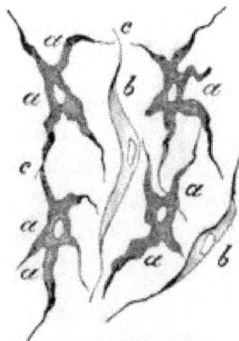

Fig. 7.

Fig. 7. Cells of the black pigment of man seen from the surface. (*Kölliker.*)

Fig. 6. Cells of the choroideal network. *a.* Pigmented cells. *b.* Pigmentless fusiform cells. *c.* Anastomoses of the pigmented cells. (*Kölliker.*)

coat of arteries. The investigations of Virchow and Donders seem to have demonstrated that the stroma under consideration belongs to the undeveloped form of elastic tissue. This tissue is composed of cells with numerous prolongations or processes, which form a network with straight and elegant meshes, as seen in the cellular tissue of other parts, but differ from ordinary cellular tissue in being pigmented. The black pigment completely lines the inner surface of the choroid

membrane as a connected cellular lamina, and as far forward
as the ora serrata, as a single layer of beautiful cells, almost
regularly hexagonal, 0.006''' to 0.008''' in diameter (Fig. 7), ·
and 0.004''' in thickness, disposed in the manner of a mosaic.
The large quantity of pigment in them allows the cell nucleus
to appear only as a clear spot in the interior ; but this nucleus
is seen on a lateral view to be situated in the outer half of the
cells, where they are poorer in pigment granules (Kölliker).
The pigment is not deposited on the vessels so much as in between them. It is more abundant immediately around the
optic nerve entrance, than in other parts of the choroid, and
is seen sometimes in a crescentic shape along the edge of the
nerve entrance. This *stratum pigmenti uveæ* lines the whole
vascular membrane from the edge of the pupil to the optic
nerve entrance. The pigment-cells are so filled with fine pigment granules, that their nuclei are seen as light, pale, pigmentless vesicular spheroids, with 1 or 2 floating nuclei. This
layer gives to the uvea the color of black silk velvet, to
which it has always been compared. In blonde persons this
pigment is but little developed. This pigment stratum is covered on its whole extent over the uvea by a structureless membrane, covered by very bright granules. After covering the
posterior surface of the iris, it covers the whole free surface of
the ciliary processes and the choroid. It is attached at the *ora
serrata retinæ*, and there it is also connected with the *membrana
limitans retinæ*. Stellwag names it the *membrana limitans uveæ*.

On its outer surface, next to the sclerotica, we find that the
stroma of the choroid retains its characteristics, only the cells
become less numerous, the fibres longer, more fine, and less
resisting to the action of reagents, forming larger meshes.
This part of the choroid is named *lamina fusca*, but inaccu-
rately so (Manz), as histology teaches us that it is not independent. However, in a certain number of instances, it appears
as a continuous membrane, and not only as a layer of loose
cellular tissue. It forms the bed in which lie the vessels and
nerves that proceed to the iris and ciliary body, and it varies

much in thickness in different subjects. It ceases at the ciliary body, and is gradually lost toward the optic nerve entrance.

Smooth muscular fibres were discovered in the choroid by Heinrich Müller and by Schweigger. They are found near the larger choroideal vessels, and are seen as bands of an opaque tissue, of about half the size of the arteries. On treating these bands with acetic acid, a crowd of longitudinal nuclei are seen, altogether similar to the fibre-cells of the ciliary muscle. Their successful investigation seems to be extremely difficult.

In recent years *nerves* have also been discovered in the choroid (H. Müller, Schweigger, Stellwag, Pope, etc.) They are numerous, and are found only in the posterior part of the globe, consisting not only of nerves of double contour, emanat-

Fig. 8.

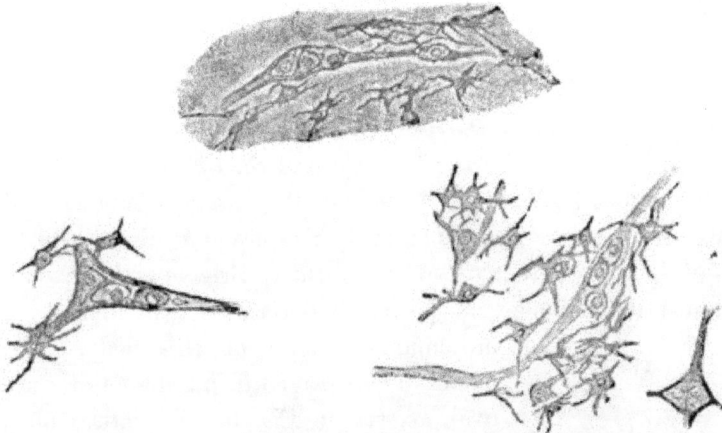

Ganglion-cells and pale nerve-fibres found in the choroideal stroma.
(*From Schweigger.*)

ing from the ciliary nerves after their passage through the sclerotic, but also of a plexus of pale fibres, the extremities of ganglion-cells. The emanations of this plexus seem to lose themselves in the walls of the vessels, and in the bands of muscular fibres described above. Schweigger says that the ganglion-cells, pale nerve-fibres, and the smooth muscular fibres, are found in the inner vascular layer of the choroid, near the *chorio-capillaris.*

The functions of the choroideal nerves are not known, but probably they possess a regulating influence on the choroideal circulation. It is not unlikely that they, in connection with the smooth muscular fibres, have an influence on the accommodative process, acting as antagonists to the ciliary muscle (*Manz*).

The inner surface of the choroid is uniform, until within 2½''' from the anterior border of the sclerotica, where we observe a dentated line, the *ora serrata*. At this point the *choriocapillaris* ceases, and the second division of the vascular membrane begins.

The corpus ciliare, which is bounded posteriorly by a dentated line, the *ora serrata retinæ*, extends forward to the canal of Schlemm, and the periphery of the iris. When viewed from behind, forward, it forms a beautiful deep-brown ring, from 2½''' to 3''' broad, and a little more narrow on the nasal side than on the temporal. It consists of the *musculus ciliaris*, of a part of the choroideal stroma which passes over into the iris stroma, and of the *processus ciliaris*.

The *musculus ciliaris* (*Tensor choroidea, ligamentum ciliare*) consists of a layer of radiating smooth muscular bundles, and near its anterior and inner part, also circular fibres, which, doubtless, are antagonists to the former. Brücke and Bowman almost at the same time positively determined the undoubted

FIG. 9.

muscular character of this body. The nuclei in these smooth muscular fibres are oval, as seen in Fig. 9. The fibres form small fasciculi, between which are found cellular tissue, bloodvessels, nerves, and ganglions. The direction of these bundles is, from before, backward, with a slight convergence at their posterior extremities toward the antero-posterior axis of the eye. The whole muscle is a rather thick muscular ring, triangular or arrow-shaped, its apex extending back as far as the *ora*

Muscular fibres from ciliary muscle. (*From Brucke.*)

serrata, whilst the thicker end proceeds

forward to the peripheral insertion of the iris, at the canal of Schlemm, into the inner wall of which it is inserted by a short, circular, white ring, which forms a portion of the canal, or venous sinus, named. It gradually increases in thickness from the apex to its insertion, and gains a thickness of $\frac{2}{3}'''$ to $\frac{1}{2}'''$, the whole *corpus ciliare* being at the same place $1'''$ thick.

At its posterior thin portion the outer layer of the stroma of the choroid divides, according to Brücke, the inner layer with

Fig. 10.

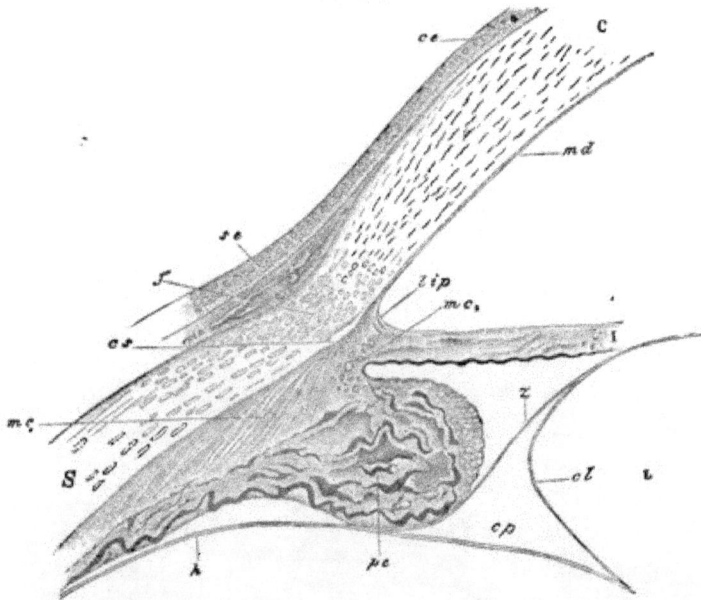

Fig. 10. Ciliary region of the human eye; section made along the antero-posterior axis of the globe. Magnified 15 diameters.

C, cornea; e e, epithelial layer of the cornea; m d, membrane of Descemet; f, union of the cornea with the sclerotica.

S, sclerotica; s e, epithelial layer of the bulb; *episc*, episcleral cellular tissue (originating from the basis of the ocular conjunctiva); e s, canal of Schlemm; $m c_1$, radiating fibres of the ciliary muscle; $m c_2$, section of the circular fibres of the ciliary muscle.

I, iris; lip, pectinate ligament of the iris; p c. section of the ciliary processes.

L, lens; c l, capsule of the lens; z, zone of Zinnius; h, hyaloid membrane; e p, Petit's canal. (*From Manz.*)

the larger vessels extending on the inner surface of the muscle, to go to the iris, with the stroma of which it becomes continu-

ous. The outer layer passes over the external surface of the
muscle in the form of fascia. At the anterior border of the
muscle this fascia takes on again some of the characteristics of
the inner layer, and with it passes into a firm network of non-
nucleated fibres of a peculiar structure, which is, as above
stated, inserted into the inner wall of the *canalis Schlemmi*, as
a short ring-formed tendon of the ciliary muscle. It projects
furthest forward at the posterior part of the inner wall of the
canal of Schlemm, which is the so-called tendon of the muscle.
In its structure we find the fine, smooth fibres, consisting of
fine, granulated, very tender and easily destructible fibre-cells,

Fig. 11.

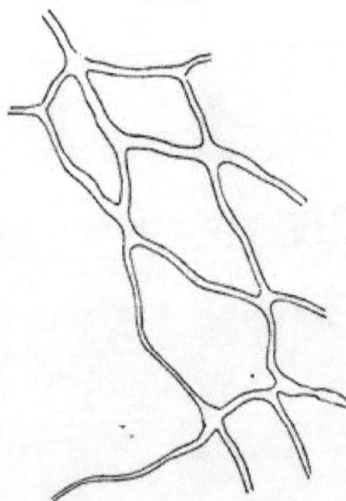

Non-nucleated fibres forming the tendon of the ciliary muscle.   (*From Brucke.*)

which are somewhat shorter and broader than the ordinary
cells. The pale-gray fibres are so tender, and so easily broken
down, that their investigation is rendered difficult.

Recent investigations seem to prove that a portion of the
muscular fibres also originate from the fibrous net of the aqueous membrane. The *inner* strata of the muscular ring extends
itself (within the limits of the ciliary processes) toward the
eye-axis, bending into the folds of the processes in such a

manner that their whole outer surface hangs over the muscle-ends, like a brush.

The circular fibres interlace freely with the longitudinal fibres, some assuming the arched form, their ends being turned backward (Stellwag). I am not aware that it has been dis-covered that these circular fibres belong to the striated class of muscular fibres, although it is quite likely that they are under the control of the will.

The inner surface of the muscle is lined by the middle layer of the choroid, composed of choroideal stroma and the larger choroideal vessels, and passes over into the stroma of the iris, with which it becomes continuous. It is somewhat firmly connected with the muscle by its external surface by a dense fascia; on its inner surface it is connected with its posterior $\frac{2}{3}$ with the *processus ciliares*, whilst its anterior $\frac{1}{3}$ turns toward the origin of the iris, and to the tendon of the ciliary muscle, so as to form, with the ciliary processes, an angular border. (See Fig. 10.)

The *third* and innermost division of the *corpus ciliare* origi-nates from the inner pigment-layer of the choroid. At the *ora serrata retinæ*, $2\frac{1}{2}'''$ to $3'''$ from the anterior border of the *scler-otica*, the membrane is seen to swell into ridges, and as these pro-ceed forward, they gradually increase in size, and constitute the peculiar structure called the ciliary processes. In the posterior half of this beautiful, dark-brown ring, the *striæ* are small, so as not to puff up the membrane much, and is called *pars non fimbriata corporis ciliaris;* the anterior half, being more puffed up, is called the *pars fimbriata*. In the latter division, the choroideal stroma is intimately connected with the plaited neck of the *hyaloid membrane*, the *zonula Zinnii*, and is quite rich in pigment, having the appearance of being covered by black velvet. These projections are from 70 to 80 in number, gradually increasing in size as they proceed forward, and are of a whitish-gray color. Each elevation begins an obtuse angle from the middle layer of the *corpus ciliare*, and proceeds for-ward toward the iris, but never coming quite in contact with it, as some assert.   It is difficult to determine the precise relations the ciliary processes sustain, in the living eye, to the lens and the iris. Their relative position varies in accordance with certain intraocular changes. They are placed to the outer and anterior edge of the lens, *but are never in contact with any part thereof.* According to some very interesting observations, recently published by Otto Becker, on the position of the *processus ciliares*, made on albinotic persons, and also on a case of *iridæmia*, the processes in no case touch the margin of the lens. He also ascertained that, during accommodation for near objects, the pupil contracting, the ciliary processes are diminished in size, and are drawn backward. The same phenomenon takes place under the action of the Calabar bean; the pupil contracts, the ciliary processes diminish in size and recede, and the eye becomes myopic. On the other hand, when the pupil dilates, either for accommodation for the far point of vision, or from the effect of atropine, the ciliary processes enlarge, and protrude forward toward the optic axis, but without coming in contact with the lens; but they lie between it and the iris, separated by a distinct space. It appears that the processes do project forward and outward, between the iris and the lens, into the posterior chamber, without touching the periphery of the lens nor the iris, as seen in the figures of Arlt, Jæger, Hasner, and Bowman. The cause assigned by Becker for this swelling of the processes, is that, when the iris is dilated, the free circulation of blood through it is somewhat interrupted, and, consequently, it is dammed up in the ciliary processes, which it swells up.

Each process has a breadth of $\frac{1}{10}'''$ to $\frac{1}{5}'''$, a height of $\frac{2}{3}'''$ to $\frac{1}{2}'''$, and a length of $\frac{4}{5}'''$ to $1\frac{2}{3}'''$.

In structure they are the same as the stroma of the choroid, only that here the stellate cells are more rare and more tender, and that here, with the exception of the base of the processes, it is but little pigmented.

The entire *corpus ciliare* is attached firmly with the sclerotica only at the tendon of the ciliary muscle, its outer surface only having a slight vascular connection from the vessels entering and leaving it through the sclerotica. With its inner surface it is connected with the retina at the *ora serrata*, and from that point forward it is inseparable from the hyaloid body.

The *corpus ciliare* has a breadth at the nasal side of $2\frac{1}{2}'''$, at the temporal $3'''$, at the posterior edge it has a thickness of $_2{_0}'''$ to $_1{_0}'''$, whilst its anterior part is $1'''$ in thickness.

The Iris.—The iris is the *third* division of the vascular membrane, the stroma of the choroid being continuous with the stroma on its posterior surface, and lines it to the margin of the pupil. It contains, however, elements quite different from the choroid. It possesses true cellular tissue, which is loose, and arranged in waving *striæ*. This connective tissue is in part radiary and in part circular, the latter arrangement predominating near its ciliary attachment. These bundles interlace with each other very freely. This connective tissue contains numerous cells, mostly spindle-shaped and stellate; more rarely round or connective tissue corpuscles, resembling the cells of the choroideal stroma, are found, being heavily pigmented, and with its fine processes they are connected into a network (Stellwag, Pilz). In dark eyes these cells possess brown or black pigment.

In the anterior layer of the iris some of the fibres of the *ligamentum pectinatum iridis* are mixed up with the cellular tissue, but they do not extend beyond the half of the breadth of the iris. Inclosed within the cellular stroma are bundles of smooth muscular fibres, bloodvessels, and nerves.

Part of the muscular fibres are circular, and form a sphincter muscle (*sphincter pupillæ*) in the form of a smooth ring, $\frac{1}{4}'''$ in breadth, and located close to the pupil, nearer the posterior than the anterior surface. Kölliker discovered another muscular ring, very small, not more than the $_3{_0}'''$ in breadth, situated nearer the anterior surface of the iris, close to the *annulus iridis minor*. The appearance of the muscles of the iris, under the microscope, is seen in Fig. 12.

The radiary fibres (*dilatator pupillæ*) do not form a continuous muscle, but extend from the ciliary border in a radiary manner, and are collected into slender fasciculi, which, as they approach the sphincter, divide, and are inserted into the circular fibres divergently, forming two bows, sometimes forming

Fig. 12.

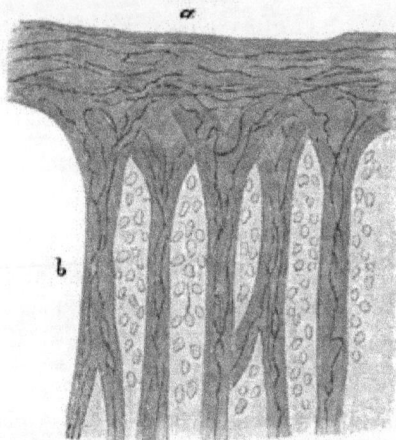

Muscle of the iris as seen under the microscope.   (*From Pilz.*)

regular arches, as shown in Fig. 12.   The fasciculi of the *dilatator pupillæ*, after reaching the circular muscle, are quite near the posterior surface of the iris, immediately in front of the uvea.

In recent years the existence of the dilatator muscle of the iris has been denied (Grünhagen, *A., Ueber Irisbewegung, Arch. f. Path. Anat. und Physiol.,* Bd. xxx, Heft 5 and 6, p. 481). Now, there is truth in this, if it is claimed that it is a continuous connected muscle, which it is not.   But, as described by Henle, as a layer of very thin muscular fibres on the posterior surface of the iris, covered and permeated by pigment-cells, which renders its discovery difficult, the claim of its existence must be insisted on.

The origin of the radiary muscular fibres has not yet been positively determined.   It is asserted by many that they originate from the *ligamentum iridis pectinatum*, and from the border of the aqueous membrane of the cornea. Kölliker believes that they arise from the circular fibrous layer on the inner wall of the *canalis Schlemmi*, from which the ciliary muscle also originates. Stellwag believes that a large portion of these fibres pass backward, with the iris stroma, vessels, and nerves, to the space between the ciliary muscle and the outer border of the ciliary processes, where they expand in a fan-shaped manner. Here the outer fibres run in a parallel manner with the ciliary muscle as far as the *ora serrata retinæ*, whilst the inner fibres turn in toward the visual axis, almost perpendicularly to the stroma connecting the ciliary processes with the zonula, in which they disappear. He claims to have preparations which demonstrate this origin of the radiary fibres of the iris. The consonance of action of the ciliary muscle and the iris during the accommodative act, would seem to point to some such intimate anatomical connection between these parts.

Anteriorly the iris is covered by a single layer of round and flattened epithelial cells, which is an immediate continuation of the epithelial covering of the posterior surface of the *membrana Descemeti*.

The posterior covering of the iris is called the *uvea*, and is continuous with the choroid and ciliary processes, covers the whole posterior surface of the iris to the pupillary edge, which it hems in, so as to be visible on the front of the pupillary margin. It consists of a thick stratum of cells, which are round, densely packed, and filled with dark pigment-molecules. Between these pigment-cells and the iris itself is a membrane, which Kölliker considers the same as the *membrana limitans choroidea;* others claim it only to be formed by the anterior union of the walls of the pigment-cells. This layer in light eyes is poor in pigment, and reflects blue or gray; in dark eyes it is richer in pigment, and gives to the eye a dark brown or black appearance. In such eyes the pigment is not confined to the posterior tapetum or layer, but is found in the stroma of the iris, in its anterior epithelium, in

the form of golden yellow or brownish granules. They are
said also to be found free between the muscular fibres and
vessels.

The manner in which the anterior surface of the iris is
formed by the *anterior elastic lamina*, or *membrana Descemeti*,
was first described by Bowman. The *membrana Descemeti*,
near the border of the cornea, begins to send off from its anterior surface fine elastic fibrils, which constantly increase in
thickness, until at the margin of the cornea the whole thickness of the membrane of Descemet is broken up into thicker
fibres and columns, which turn over on the anterior surface of
the iris to form the anterior fibrous membrane and pillars, or
the so-called *ligamentum iridis pectinatum* (Bowman and Kölliker). It will thus be perceived that the iris has no peripheral

FIG. 13.

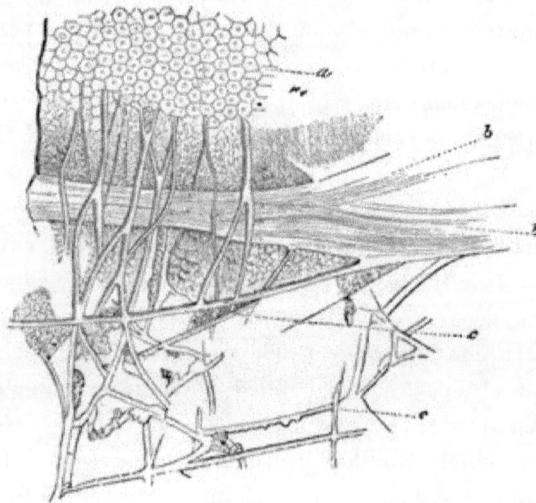

FIG. 13. Union of the Membrana Descemeti with the ligamentum iridis pectinatum.
a. Membrane of Descemet. *b*. Tendinous ring or wall of canal of Schlemm. *c*. Ligamentum pectinatum expanded on the anterior surface of the iris. (*From Manz.*)

attachment, but is immediately continuous all around with the
membrana Descemeti. All the fibres of the membrane of Descemet at the margin of the cornea, do not turn in to be ex-

panded on the anterior surface of the iris, a part of them only
assuming this direction; another portion of them pass into the
anterior two-thirds of the wall of the canal of Schlemm. The
nature of these fibres has not been settled. They have been
considered as connective tissue, as serous fibres, in part elastic,
and partly cellular, and as intermediate between the elastic and
connective tissues (Henle, Luschka, Bowman, Kölliker, Pilz).

The vascular system of the iris is intimately connected with
that of the ciliary body and the choroid. We are indebted to
Leber for recent clear elucidations on the circulation of the
iris, ciliary body, and the choroid.

The *iris* is supplied with blood from the *arteriæ ciliares posteriores longæ*, the *arteriæ ciliares anteriores*, and the *arteriæ
ciliares posticæ breves*. The posterior long ciliary arteries are
two in number, one passing forward on the inner side of the
eye, and the other on the outer side. They perforate the
sclerotica a little anterior to the posterior short ciliary arteries, passing it so obliquely that there generally intervenes a
space of several lines between the points of entrance and exit.
They run forward between the sclerotica and the choroid, enveloped by the *lamina fusca*, to the posterior border of the
ciliary muscle, where they divide into two branches, which
rapidly divide into smaller branches.

The *arteriæ ciliares anteriores* are branches from the arteries of
the four straight muscles: (exceptionally one arises from a palpebral artery.) They pass through the tendons of the muscles to
the sclerotica, and run toward the cornea, near the margin of
which they perforate the sclerotica somewhat directly from
without inward to the ciliary muscle, where they join the
branches of the long posterior ciliary arteries to form the *circulus arteriosus iridis major*. This is a complete, closed, vascular
ring, in some places being double, and in others threefold. It
is situated in the ciliary muscle, in its anterior border near the
periphery of the iris. From this arterial circle, arteries pass
forward to the iris, inward to the ciliary processes, and backward to the ciliary muscle and the choroid. The arteries

4

passing to the iris and to the ciliary processes generally arise from one common trunk, which divides into two branches, one proceeding to the iris and the other to the ciliary processes.

Besides the large arterial circle of the iris, the branches of the posterior short ciliary arteries, and the anterior ciliary arteries, form an imperfect circle or row of anastomoses further back in the ciliary muscle, from which proceed the arteries of the ciliary muscle, and the arteries passing back to the choroid, which are generally from ten to twelve in number. It will thus be perceived that, according to Leber, the ciliary body and the iris are much more independent of the short posterior ciliary arteries than has ordinarily been supposed. The ciliary muscle possesses a very large number of small vessels, which spread within it in an arborescent manner, and form a quite close capillary network, the mass of which follows the direction of the radiary muscular fibres in a parallel manner on its external surface. On the internal surface it is more irregular. The arteries of the ciliary processes and of the iris generally originate in one common trunk from the large arterial circle of the iris. Those of the former rapidly pass into a mass of fine twigs, which anastomose freely, and pass to the free border of the processes in a curved manner, where they terminate in the commencement of the veins, which seem larger than the arteries.

The arteries of the iris form in it a loose capillary network, especially on the *sphincter pupillæ*. Some branches form at a certain distance from the pupillary border the well-known *circulus arteriosus iridis minor*, which is, however, not a perfect circle. At the pupillary border the fine arterial branches pass in a looped manner into the beginning of the veins.

Veins of the Corpus Ciliare and the Iris.—The blood from the vascular membrane is carried off through the *vasa vorticosa*, the *venæ ciliares posteriores*, and the *venæ ciliares anteriores*. The *venæ ciliares posteriores breves* are small, and, according to Leber, receive blood *only* from the *sclerotica;* and none from the choroid. The *vasa vorticosa* have been described. The long posterior ciliary veins, Leber says, cannot be found, as delineated in works on the anatomy of the eye. Brücke says they are somewhat less in size than the arteries, and accompany them.

The blood from the vascular membrane, then, is carried off by the *vasa vorticosa*, with the exception of a small portion which is conveyed through the *venæ ciliares anticæ*, being much less in quantity than the *arteriæ ciliares anticæ* carry into the eye. The veins of the iris run back between the ciliary processes, to empty into the *venæ vorticosæ*. Leber never saw any of the veins of the iris empty directly into the *canalis Schlemmii*, as has generally been taught. They are joined by the veins from the ciliary processes, and by some also from the ciliary muscle. The veins of the ciliary processes arise from the vascular network of that structure, a larger vein generally running along the free border of the process. In the space between the processes, the veins returning from the iris run back, and, with the veins of the processes, form a venous network, which lies immediately beneath the inner surface of the ciliary body, the trunk passing over the smooth part of the *corpus ciliare* to the choroid, and at the border of the latter membrane it passes to the outer layer thereof. The veins of the iris and the ciliary processes pass inside of the ciliary muscle, and are clear of it, whilst the arteries running to these parts all have to pass through the muscle. Consequently, during the contraction of the ciliary muscle, for accommodation for near objects, the compression of the arteries lessens the amount of blood thrown into the iris and *processus ciliares*, whilst the veins are left free to disgorge themselves. It has been the common opinion that, during accommodation for near objects, the ciliary processes become engorged, which, from the arrangement of the vessels in the parts concerned, is not possible. In the observations of Otto Becker, recently made on albinotic persons, he observed that, during the contraction of the ciliary muscle for adjustment for the near point of vision, the ciliary processes receded and diminished, and that, in adjustment for the far point of vision, they increased in size.

As regards the *venæ ciliares anticæ*—whether some derive their blood immediately from the so-called *canalis Schlemmii* or from the ciliary muscle, has been a disputed matter among authors. It has generally been believed that some of the veins of the iris empty into the canal of Schlemm, and that the *venæ ciliares anticæ* convey it from thence out of the eye. Brücke, who is very high authority, takes this view. In Note No. 27, annexed to his Treatise on the Eyeball, he says that this is the only instance in which he did not follow the results of his own investigations, but that he followed Arnold and Retzius, and that he could never succeed in tracing any veins from the iris into the *canalis Schlemmii*, nor inject any, from the canal outward, with mercury. Leber could never trace any of the veins from the iris into this venous sinus. As regards the *canalis Schlemmii* itself, it has been considered as a venous *sinus*, regularly round, and belonging to the system of the *venæ ciliares anticæ*. The anterior two-thirds of its wall consists of elastic tissue, the fibres of which originate from the *membrana Descemetii*, and the posterior third is a lamella of tendinous substance, which springs from the sclerotica, and is identical with it in texture. Its posterior wall is much thicker than its anterior. At the point where the elastic and tendinous tissues unite, the ciliary muscle and the radiary fibres of the iris arise from a common origin. From the anterior elastic portion of the wall of this canal the elastic fibres start that form the *ligamentum iridis pectinatum*. Its horizontal diameter, according to E. Jaeger, is 0'''.3058, and its horizontal plane, from side to side, measures 12'''.5666. It is situated in the wall of the sclerotica, and much nearer its anterior border, quite close to the corneal border.

Leber asserts that it is not a circular canal, as has been taught, but that it is a circular *venous network*, in the innermost layer of the sclerotica, and immediately external to the insertion of the ciliary muscle. He thinks it ought to be denominated *plexus ciliares venosus*. In this plexus of veins there are, at certain points, 6 to 7 veins of nearly the same thickness, which, by free intercommunication, form a close network of veins. In other parts of the circumference one or two large veins are accompanied by a few smaller, which freely communicate with each other. This circular plexus seems, at first view, like a canal, but close investigation, Leber says, will prove that it is not. The loose connective tissue around the veins will permit a small probe to force a passage. In some places small islands are seen, formed by a vein running out and returning again to the main trunk.

The anterior portion of the ciliary muscle sends forward and outward numerous small venous trunks, which perforate the sclerotica as *venæ ciliares anticæ*. Other small veins enter the posterior wall of the *canalis Schlemmii*, or, rather, *plexus ciliares venosus*, and at each point of the entrance of a vein from the ciliary muscle, several small branches perforate the sclerotica, to form anterior ciliary veins. But all the blood passing off through the *venæ ciliares anticæ* is much less than the amount thrown into the eye through the *arteriæ ciliares anteriores*, the larger portion escaping through the *venæ vorticosæ*. Hence, it seems, according to Leber, that the so-called *canal of Schlemm* receives blood only from the veins of the anterior part of the ciliary muscle; that it has no direct communication with the anterior ciliary veins, but through its elongations toward the ciliary muscle; and that it must be considered a venous reservoir for the ciliary muscle, into which blood can escape during its contraction, and return again when the contraction ceases.

The iris and ciliary muscle are supplied with *nerves* mostly from twigs from the *trigeminus* and the *oculo-motorius*. Some, however, also are derived from the *sympathetic* and *abducens*. They enter the eye mostly as the *nervi ciliares breves* from the ciliary ganglion, and perforate the sclerotica around the optic nerve entrance, and pass forward through the *lamina fusca*, to be distributed in the ciliary muscle, iris, and cornea. There are two (sometimes only one) nervous twigs, called *nervi ciliares longæ*, derived from the *ramus nasa ciliares*, the first branch of

the *trigeminus*. They perforate the sclerotica immediately be
hind the insertion of the *musculus trochlearis*, and pass forward to the ciliary muscle, in which they ramify with the
short ciliary nerves. In recent years it has been discovered
that many of the nervous branches of the ciliary muscle are
surrounded by ganglion-cells, or by small agglomerations of
these cells (H. Müller, Liebreich, Krause, Manz). They con-
tribute to the formation of the nervous network in the ciliary
muscle. From this network the iris is provided. They run
somewhat like the bloodvessels, and, like them, form anastomoses, arches, and circles, one of which corresponds with the
circulus arteriosus minor, and another circle is formed a little
more outward. These nerves being motor nerves, it is probable that they lose themselves in the muscular structure.

We know that the third pair of nerves sends twigs to the
ciliary muscle and iris, as paralysis of that nerve causes paralysis of accommodation. We know also that there are both
sympathetic and cerebro-spinal nerves in the eye, from the
fact that the antagonistic action of opium and of belladonna
are witnessed, the former causing contraction of the pupil and
spasm of the accommodation, whilst atropin causes dilatation
of the pupil and paralysis of the accommodation, the cerebro-
spinal nerves acting on the circular fibres, and the sympathetic
on the radiary fibres (Graefe).

The iris, considered as a whole, is a quite soft, loose tissue,
highly yielding; it can be stretched out more than half without tearing. It has but one border, the pupillary border, the
ligamentum iridis pectinatum being continuous with the *membrana Descemetii* anteriorly, whilst posteriorly its stroma is
continuous with the choroidal stroma. The *pupil* is located
¼''' nearer the nasal than the temporal side, and varies in
diameter from 1''' to 3'''. According to Jaeger the iris is
thickest in the middle, 0'''.45; at the ciliary and pupillary
borders it is 0'''.30. Its color depends on the number and
arrangement of the pigment-cells.

The anterior surface of the iris is divided into two zones by a zigzag line, of which the outer iris circle is the larger. In light-colored eyes the small circle is darker than the large, whilst in dark-colored eyes it is often the lighter in color. The color varies from gray brown to dark brown. The direction of the fibres is radial in this zone, which, the nearer they approach the small zone, the more they separate, and leave black, oblong spaces between them, which often pass over the boundary line between the zones. The inner circle proceeds from below and behind the larger circle, and at the pupillary border there is a raised ridge or rim derived from the *uvea*. The color of this circle is generally gray or bright gray, and with a dark larger circle it is rust-colored. The degree of prominence of the anterior surface of the iris varies very much. During accommodation for the near point of vision, the small circle projects forward, and the ciliary border is drawn back. In accommodation for the far point of vision the reverse of this takes place. When the ciliary processes are congested they doubtless push the larger circle of the iris forward. The pupillary border, doubtless, is in contact with the anterior capsule of the lens, having only the moisture of the aqueous humor between them. The posterior chamber is much smaller than the anterior, yet a small quantity of the aqueous humor is always between the posterior wall of the iris and the ciliary processes, and the capsule of the lens.

The Retina (*Tunica Retina*).

The retina extends from the entrance of the optic nerve, being in part in continuous connection with it, to within $\frac{5}{4}'''$ of the *corpus ciliare*, near the *ora serrata retinæ*, where its proper nervous character ceases, and where it is firmly connected with the choroid. At the point named there is a slight puffing of the membrane, and when the retina is torn loose for investigation, a finely serrated edge is left, and hence the name of *ora serrata*. The delicate gray membrane, which has a thickness of $0'''.018$ to $0'''.02$, and lies on the inner

surface of the *membrana limitans uveæ*, and is closely connected
with it, called the *pars ciliares retinæ*, is a continuation of the
connective tissue fibres of the retina, and not an epithelial membrane, as taught by Hanover, Pilz, and others.   The retina is
an entirely transparent substance, and becomes visible by ca-
daveric changes, or by detachment from the choroid, or when
subjected to a hardening solution (Ritter).

It has a thickness in the middle of 0′′′.1, whilst toward the
ora serrata it is thinned out to 0′′′.04.   It has a surface of
about 300 square lines.   Externally it is in close contact with
the vascular membrane, and internally with the hyaloid covering of the vitreous body, over the convex surface of which it
is expanded.

In its structure the retina is a very delicate, complicated
tissue, on which histologists by no means agree.   A history of
the various opinions of microscopists on this subject would be
too voluminous for these pages, and perhaps without much
advantage.   Recent histologists, however, all seem to agree
that the late Heinrich Müller was the great pioneer in this
field of labor, and that he laid the foundation deep and solid
for other investigators.   It seems that Carl Ritter (*Die Structur der Retina dargestellt nach Untersuchungen über das Walfischauge; L'Anatomie de la Retine*, written for Wecker's *Études
Ophthalmologiques*, etc.) has been most successful in this part
of the vast histological domain, and what follows on this subject is mainly drawn from his labors.

Notwithstanding the variations in thickness of the retina,
the following layers can be distinctly traced from without inward:

1. The layer of rods and cones (Stratum bacillorum, Membrana fucoli).

2. The granular layer (outer granular layer).

3. The outer fibrous layer (the intermediate granular layer).

4. The layer of granule cells (inner granular layer).

5. The inner fibrous layer (layer of gray nervous substance,
fine granular layer).

6. The layer of ganglion-cells.

7. The expansion of the optic nerve fibres.

8. The limitary membrane (Membrana limitans retinæ).

The *external limitary membrane* of Max Schultze, between the granular layer and the rods and cones, is not in reality a membrane, and will be considered hereafter. Mixed in with the nervous structure of the retina there is that kind of connective tissue that Virchow has discovered in the brain, called *neuroglia*. This connective tissue extends from the internal limitary membrane to the *inner* surface of the rod and cone layer.

The Nervous Tissues of the Retina.—In the retina are found the terminations of the optic nerve fibres in connection with the sensorial apparatus on which the luminous impressions are made. This connection of the apparatus on which the luminous impressions are made and the optic nerve fibres which conduct those impressions to the brain, takes place through the medium of the ganglion-cells. There are numerous sensitive points to each conducting fibre, many rods and cones being in immediate communication with it, through the medium of fine nerve-fibres (*fibres of Müller*), which are in connection with all the nervous layers, which layers run parallel with the retinal plane. Ritter says that the true nervous layers consist of

1. The layer of rods and cones.

2. The granular layer (external granular layer).

3. The layer of granule-cells (internal granular layer).

4. Layer of ganglion-cells.

5. Layer of nervous fibres.

The outer and inner fibrous layers are simply layers of conducting fibres, without possessing any peculiar structure or function.

The Layer of Rods and Cones (*Membrana Jacobi*).—This is the outermost layer of the retina, and consists of two kinds of elements,—the *rods* and the *cones*. The outer surface of this layer rests against the pigment-layer of the choroid. The inner surface is connected with the granular layer. It consists of regularly arranged rods and cones, placed vertically, in palisade form, and packed closely together. In some parts of the retina, the rods, and in other portions, the cones, preponderate. At the *macula lutea* there are no *rods*, but the layer is entirely made up of *cones*. At the border of the yellow spot the rods already predominate, at the middle of the retina there are still less cones in proportion to the rods, and further toward the periphery of the retina, near the *ora serrata*, the disproportion at the expense of the cones is still more marked. (See Fig. 14.)

The rods and cones form a single layer, with a thickness at the centre of the retina of $0'''.036$, further forward only $0'''.030$, and nearer the periphery it diminishes to $0'''.028$ (Kölliker).

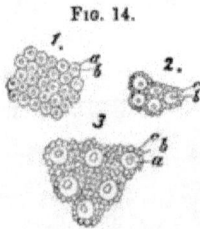

Fig. 14.

The Rods.—The true character of the rods and cones remains a matter of dispute. The figures of rods and cones given by different authors vary very much. As heretofore stated, it seems that Ritter has been eminently successful in his investigations of the retina, being the first author to present clearly the character and termination of Müller's fibres. The figures given by Pilz, Kölliker, H. Müller, Fick, and nearly all who have written on the retina, represent the rods or the cones, or both, as terminating with a broad triangular base on

Bacillar layer seen from without. 1, at the yellow spot (only cones); 2, at the boundary of this spot; 3, from the middle of the retina; *a*, the cones, or the spaces corresponding to them; *b*, rods of the cones, the terminal surface of which is often situated deeper than the ends of the proper rods, *c*. Magnified 350 times. (*From Kölliker.*)

the *membrana limitans internæ*. The subject is an extremely difficult one to investigate. The idea of a cone or rod, on which the luminous impressions are made, running through all the layers of the retina, to terminate with a broad base in the *membrana limitans interna*, instead of terminating in a ganglion-cell, or an optic nerve fibre, is a physiological incomprehensibility. Ritter, by his investigations on the eyes of the whale, has arrived at the conclusion that every rod and every cone communicates, by means of the fibres of Müller, with a ganglion-cell; and that the optic nerve fibres all terminate in ganglion-cells is a known fact. This offers us a connected, comprehensible view of the different elements of the retina. The triangular expansions seen in vertical sections of the retina, terminating by a broad base on the *membrana limitans interna*, are the connective tissue of the retina, which starts with a broad base from the limitary membrane, but narrows so as to pass between the ganglion-cells, in a manner to be hereafter described.

The *rods*, says Ritter, are complete cylinders, which measure in the adult 0.05 mm. in length, with a thickness of 0.003 mm. In the fresh state, each one presents a yellow reflection, and has four well-defined surfaces, of which the two largest, which are exactly parallel, are cut by the two smallest at nearly right angles. At the inner extremity of each rod there is sometimes seen a fine filament, hardly perceptible.

This filament is always in communication, in its course, with a granule of the granular layer. This filament is best seen after the retina has been subjected to some hardening process; but even in the perfectly fresh state, a faint line may be seen running upward and downward, to terminate in the filament at its ends. The outer extremity of the rod is dilated like a club; the inner more pointed extremity forms a

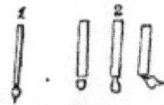

Fig 15.

Rods in the fresh state. 1. Of man. 2. Of the duck. Magnified 300 diameters. (*From Ritter.*)

dehiscence. Under careful examination it is discovered that the rod really has an opening or canal, and that the filament runs through its middle to the outer extremity, where it terminates in a rounded enlargement. In man the rods then are composed of an enveloping membrane and a central filament.

The Cones.—The cones are composed of a middle, large portion, which is granulated, and of two appendages, of which one projects outward and the other inward. The internal appendage is continued by an elongated thread, which contains in its course several granules of the granular layer; the external appendage ends by a rounded extremity. The length of the cone in the adult is 0.03 mm. to 0.04 mm. Its greatest thickness is 0.006 mm.; its appendages only 0.005 mm. Under the influence of hardening substances a globule is discovered toward the inner extremity of its middle portion, which is generally considered as a nucleus. Ritter believes it to be only a dilatation of the central filament. The external appendage or filament is not continuous with the surface of the cone, but penetrates the internal appendage, which it completely fills. From thence it is prolonged toward the internal extremity of the middle part of the cone, where it sometimes forms a globulous swelling, and sometimes terminates simply, without change of diameter.

Fig. 16.

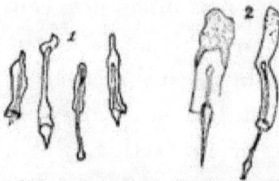

Rods hardened in chromic acid. 1. Rods of man, showing the central filaments. 2. Rods of the frog, with central filament and a medullary substance. Magnified 300 diameters. (*Ritter*.)

Fig. 17.

Cone from an adult human eye seen with the central filament. (*Ritter*.)

As regards the difference between the rods and cones, Ritter says it is only necessary to compare the composition of the rods with that of the cones to prove that no important difference exists in their character. It is only in man that the cones alone are found in the yellow spot. The importance of the cone is diminished since the discovery of the central fibre; for all the physiological value of the rods and cones rests upon the existence of this filament, of which the external extremity constitutes the point explained above. The rods and cones are only two forms of the same element. Their enveloping membrane approaches near to the cellular tissue, and it is possible that there exists between it and the arches of connective tissue (that has been traced up to the inner surface of the layer of rods and cones), a connection, which, however, remains to be proved.

The distribution of the rods and cones is such in man that in the *macula lutea* cones only exist, and they also predominate in that neighborhood, but diminish toward the periphery of the retina. In animals, no general law of distribution of these elements exists. Toward the periphery of the retina the rods and cones diminish in length. The internal, gaping extremity of the rods and the arches, most external of the cellular tissue, touch, which arrangement has given rise to the belief of the existence of a *membrana limitans externa*.

The above opinion of Ritter, that no important difference exists between the rods and cones, does not seem probable, for several reasons. It is true that it is a settled matter, that both the rods and cones are nervous elements, and that both receive luminous impressions, as the cones in the *macula lutea* exclusively prevail in man, where vision is most acute. In some animals, on the other hand, rods only are found in the retina, and yet, undeniably, such animals can see. This proves that both elements are susceptible to the impressions of light. Yet there are important anatomical and functional differences. (See Max Schultze, *Zur Anatomie und Physiologie der Retina*, Bonn, 1866.) Some of the differences between the rods and cones, very briefly enumerated, are the difference in size and form. The filaments proceeding from the inner extremity of each rod and cone also differ, those from the latter being thicker, and can be traced inward further than the former, which is very fine, and often ends by an enlargement. Each rod and each cone has a granule in connection with it, those of the cones being considerably larger than the rod granules. In many animals the cones are wholly wanting, and we *always* find this to be the case in the retinæ of such animals as live in darkness, as the bat, the mole, the mouse, and many others. Birds with acute vision have the retina plentifully supplied with cones, as much so as the retina of man. Those birds who prefer twilight, as the owl, have but very few and small cones in the retina. There is a peculiarity connected with the cones of birds. Each cone has a powerfully-refracting globule of an intensely yellow or red color connected with its extremity, through which the light

has to pass. Again, in the owl species the few cones found in
the retina have pale yellow, or colorless globules, with the red
ones entirely wanting. In certain reptiles, as the lizard and
the snake, cones only are found in the retina. This arrangement of the rods and cones is found throughout the animal
kingdom, as far as investigated. Where only imperfect vision
is needed, and no distinct perception of colors is required, and
a supply of *quantitative* light only demanded, the rods are found
at the expense or to the exclusion of the cones. On the other
hand, wherever we find acuteness of vision, with a nice distinction of colors, there we find the cones largely or exclusively
prevailing. To perfect vision three things are essential: the
perception of light, the perception of colors, and the conception
of space (Raumsinn). The first functions may be performed
by the rods alone; the cones, evidently, are connected with the
second function, and, perhaps, with the third, also.

The Granular Layer.—This lamina has generally been divided into the outer granular layer, the intermediate granular

Fig. 18.

Vertical section of the human retina near the yellow spot. Magnified 300 diameters.

. 1. Layer of rods and cones.
2, 3, 4. Nuclear layers.
2. Layer of granules.
3. Intergranular layer.
4. Layer of granule-cells.
5. Fibrous layer.
6. Layer of ganglion-cells.
7. Membrana limitans.
(*From Ritter.*)

layer, and the inner granular layer. Ritter has discovered that
the outer and inner layers are different in structure, whilst the
intermediate granular layer is wholly composed of the (nerve)
fibres of Müller, and of connective tissue, and could, with more

propriety, be named the "external fibrous layer," as suggested
by Max Schultze. This lamina separates the external and internal granular layers, the former consisting of granules, and
the latter forming a layer of cells. The whole granular layer
(including the three layers just named) has a thickness in the
central parts of the retina of 0.75 mm. Of this the external
granular layer has a thickness of 0.35 mm., the internal lamina
of 0.18 mm., and the intermediate 0.22 mm. Near the ora serrata
the entire layer diminishes to about one-fifth of the above thickness. The granular layer constitutes more than one-third of
the thickness of the retina. The granules are round or ellipsoid,
and have a diameter of 0.005 mm. to 0.01 mm. Some have a
depression on the surface (see Fig. 19), looking toward the
observer. In certain animals the round granules predominate,
as in the lamb; in other animals the ellipsoid predominate, as
in the calf; whilst in man, both exist in about the same proportion. The ellipsoid have their long diameter vertical, as

FIG. 19.

FIG. 19. Granules or nuclei. 1. Filament of the granules. 2. Transverse striæ of the
granules of the lamb. 3. The same in man. 4. The same in the calf. The right granule
in 3 shows the central depression. (*From Ritter.*)

regards the retinal plane. In 1864, Henle (*Nachrichten von der
Königlichen Gesellschaft der Wissenschaften und der G. A. Uni-
versität zu Göttingen*) discovered the transverse striæ in the
granules. He describes the granules as ellipsoids, and that
each one has three dark transverse striæ surrounding it, and
running parallel with the retinal plane. Ritter subsequently
discovered that the round and the ellipsoidal granules are
about equal in number, and that the former have, as a general
thing, two transverse striæ, whilst the latter have three. They
have a breadth of 0.001 mm., and the distance between them is

0.15 mm. These striæ disappear in a few hours after death.
The longest period that Ritter could discern them was seventeen
hours after dissolution. These striæ are found only in the mammalia. A few hours after death the borders of the striæ become
less distinct, and finally they disappear, leaving only a small
point, thus giving to the granule its dotted or granulated
aspect. These striæ can be retained several days, by a weak
solution of chromic acid, or by diluted alcohol. All efforts
made, hitherto, to determine the difference between these striæ
and the rest of the granule-body have failed. Each grain is,
likely, composed of two different substances, superimposed in
layers. The outer granules, having the depressions on which
the rods and cones rest, and which are firmly connected with
the fibre of Müller (as also the innermost granules, which have
also a firm connection with the same filaments), do not seem to
possess the transverse striæ.

These granules are contained within the fibre of Müller;
that is, the axial fibres, or central filament, as it leaves the rods
and cones, and proceeds a certain distance, expands, and em-
braces with its walls from two to five of the granules. In the
central parts of the retina the granules are more abundant than
near the periphery. The fibre of Müller consists of a very
delicate membrane, which incloses these granules.

These fibres traverse the inter-granular layer perpendicularly
in their course, where they are interlaced by bundles of cellular fibres. In this lamina two of the fibres of Müller are sometimes seen to run into each other to form one filament. The
innermost layer of the granular lamina is the thinnest of the
three granular layers, and, until Ritter's investigations, its
cells were considered the same in character as the granular
bodies of the external layer. They are cells (Fig. 21) of
0.01 mm. in diameter, and are round and polyhedrian in form.
In their fresh state they are entirely transparent, and have
the appearance of small vesicles. Within the cell there is a
finely granulated substance, and a large, round nucleus, with
distinct outlines, containing a nucleolus of 0.006 mm. diameter.

From the angles of each cell a process is given off—each cell
sending off two or three—one precisely in the internal pole of

Fig. 20.

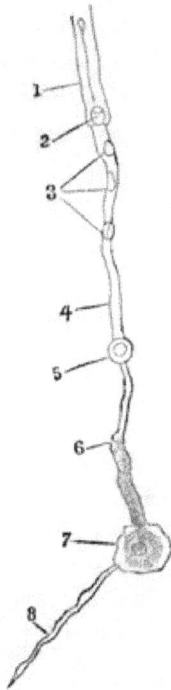

Not copied from observation, but
intended merely to show the connection of the retinal layers.
1. Rod, with the axial fibre.
2. Rod granule.
3. Fibre of Müller, inclosing
   granules of the granular
   layer.
4. Intermediate granular layer.
5. Granule cell.
6. Fibrous layer.
7. Ganglion cell.
8. Optic nerve fibre.

the cell, and the others from its external surface. The internal fibre enters the fibrous layer, and the external penetrates

Fig. 21.

Cells of the granular layer with filaments. (*Ritter.*)

the inter-granular layer. These fibres are generally denominated the *radiary fibres*. Ritter prefers naming them the fibres

of Müller, being nerve-filaments, and they must be distinguished from the connective tissue system of fibres, which will soon be described.

Two fibres originate from each cell externally, or, when only one is given off, it soon divides into branches, variable in number, which proceed toward the granules. The internal fibre, after a short perpendicular direction, runs in various directions before reaching its destination in a ganglion-cell.

The Layer of Ganglion-Cells.—The multipolar nerve-cells, constituting this lamina, have the same character as those of the brain, and have a diameter of $0'''.004$ to $0'''.016$. They are very finely granulated, and possess a nucleus of $0'''.003$ to $0'''.005$, with a distinct nucleolus. They form an unequal lamina,—in the centre of the retina being composed of eight layers of cells, the number diminishing until near the ora serrata, where they no longer form a continuous layer, but occur quite isolated. Under the influence of hardening substances the granulated contents may be removed from the cellular membrane, which is found to be a vitreous membrane, very delicate, which shows itself in ruptured cells, and in pieces prepared by tearing, in the form of small isolated scales. It is rarely that a cell contains two nuclei. The size of a cell will point out the portion of the retina to which it belongs. The smaller cells occupy the centre of the retina, and the larger are found in the periphery. This disposition of the ganglion-cells is observed in all animals; the retinal ganglion-cells have a direct relation to the size of the ganglion-cells of the brain. The same relation exists between the length and the breadth of the fibres of the retina and the ganglionic fibres of the brain, which are identical in the same animal.

Externally, the ganglion-cells give off processes or fibres, in variable number, from two to twenty-five, according to Ritter. Internally only one filament is given off, which soon becomes continuous with an optic nerve fibre. Those given off externally plunge into the fibrous lamina. The cells in the centre of the retina send off less filaments than those near the periphery. The smallest processes have a breadth of 0.002 mm., whilst the largest have a thickness half the size of the cell. These processes originate in the cell, and for some distance in the course of the fibre the granulated substance that fills the cells can be detected. These divide into a certain number of branches, in their course to the granule-cells, and in the fibrous layer they form a brush by divergent ramifications, which sometimes cross like the connective tissue fibres, with the distinction, however, that they never anastomose like the latter. These bifurcations are less common in the central

FIG. 22.

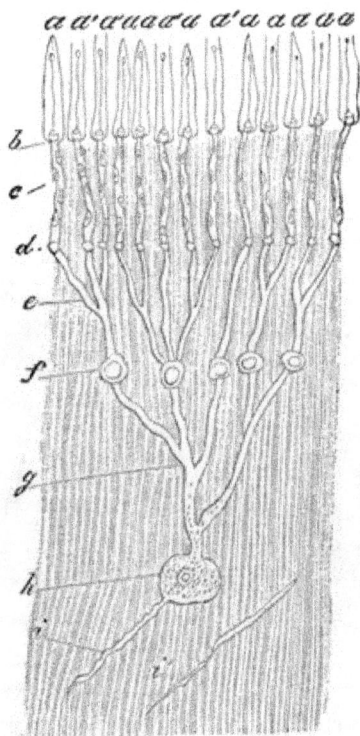

Sketch showing the connection between the various nervous elements of the retina, as taught by Ritter.—*a, a, a, a.* Cones. *a, a, a, a, a.* Rods. *b.* Rod and cone granules. *c.* Granules. *d.* Inner layer of granules. *e.* Progress of the fibres of Müller through the intermediate granular layer, many uniting in one granule-cell, *f. g.* Further progress inward of the fibres of Müller through the fibrous lamina, many uniting in one ganglion-cell, *h. i, i.* Optic nerve fibres. (Strict anatomical correctness is not claimed for this sketch, being merely illustrative.)

parts of the retina than near the periphery, which accounts for the fact that the fibrous lamina is more striated perpendicular to its direction than near the ora serrata. The fibrous lamina is made up of two systems of fibres; the one consisting of cellular connective fibres, and the other is formed by the external prolongations of the ganglion-cells. These two systems of fibres interlace, crossing each other in such a manner as to form a complicated network, that can only be unravelled by the skilful microscopist, and then only by untiring patience.

On the internal surface of each ganglion-cell but one prolongation is given off, consisting of a pale fibre of 0.0025 mm. breadth, which, soon after leaving the cell, dilates into varicose expansions with considerable regularity. They are now recognized as optic nerve fibres, running the same course, and having the same characteristics. This then traces a nervous connection between the rods and cones and the fibres of the optic nerve.

It will be perceived that this is quite a modification of the

FIG. 23.

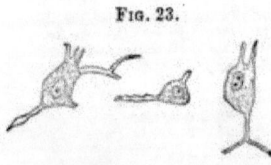

Ganglion-cells from the retina of man. Magnified 300 diameters. (*From Ritter.*)

radiary fibre system of Müller, as taught in text-books heretofore. Ritter first successfully traced the true character of the fibres of Müller in the retina of the whale (*Balæna mysticetus*). He has certainly made considerable progress in pursuing the work begun by the great Heinrich Müller, and the anatomy of the retina begins to emerge from the unsatisfactory cloudiness that previously enveloped it.

The Lamina of Optic Nerve Fibres.—This forms the innermost layer of the nervous elements of the retina, and lies on the membrana limitans. The optic nerve, from its commissure to the eye, has a structure somewhat similar to an ordinary nerve (see description of this nerve), and its varicose, dark-bordered fibres are surrounded by an ordinary neurilemma. The nerve loses its sheath in the sclerotica, and on the inner surface of the latter the neurilemma of the fibres also terminates, where

it is connected with the *lamina cribrosa*, so that in their further
progress within the eye, the nerve-tubules are divested of their
connective tissue surroundings (Kölliker). They are pale nerve-
fibres, and have a diameter of $0'''.0005$ to $0'''.001$, which, from
want of granules, are strongly refractive, and have, at least in
the dead eye, spindle-shaped varicosities.

They unite in laterally compressed fasciculi of various sizes,
which anastomose mostly by sharp angles, leaving, posteriorly,
small, and nearer the periphery, larger interspaces, which are
filled up by cellular tissue, which forms a considerable portion
of this lamina. These fibres are identical with the pale fibres
of the brain (Kölliker, Ritter). They radiate in all directions
from the *papilla nervi optici*, and form a nervous expansion,
which extends as far as the ora serrata, but is not a continuous
layer near the latter point, the fibres being only found at intervals. The thickness of the layer of optic fibres is $0'''.090$ close
to the entrance of the optic nerve; more anteriorly, $0'''.028$ to
$0'''.036$; quite in front, $0'''.002$; at the bottom of the eye,
$0'''.036$; two lines external to the yellow spot, $0'''.006$ to $0'''.008$
(Kölliker). This rapid diminution of the optic nerve fibre layer
arises from the fact that the fibres are lost in the ganglion-
cells, or, more properly speaking, in a histological sense, they
originate in those cells. It is then easily understood why the
lamina rapidly diminishes in localities where there are many
cells agglomerated. For this reason, the lamina almost disappears in the central part of the retina. At the *macula lutea*
only a small part of the optic fibres proceed directly to the
inner end; much the larger portion which are destined for the
lateral parts of the spot, describe a series of curves, which take
sweeps as they advance forward. At the yellow spot itself,
these fibres lose themselves in its deeper portion, among the
ganglion-cells, so that here there is no superficial layer of optic
fibres; the nerve-fibres of this spot are, probably, the processes
of the ganglion-cells (Kölliker).

The *membrana limitans retinæ*, on the inner side of the expansion of the optic fibres, has a thickness $0'''.0005$, and ex-

tends forward as far as the ora serrata; and some histologists assert that its connective fibres extend further forward on the inner surface of the *membrana limitans uvcæ*, under the name of *pars ciliares retinæ*. Heinrich Müller, Kölliker, and Pilz, consider it to belong, in structure, to the vitreous membranes. Max Schultze declares it is composed of cellular tissue, and that it is formed by the expansion of the radiary fibres. More recently, Carl Ritter, by his investigations on the retina of the whale, has demonstrated that it consists of cellular tissue.

The cellular tissue of the retina, says Ritter, belongs to a variety of connective tissue which Virchow discovered in the brain, and named *neuroglia*. In the mammalia, the utmost extent to which the retinal cellular tissue can be isolated is a fibre-cell, *i. e.*, a fusiform cell with two elongated extremities, and which contains a rounded nucleus. This nucleus measures 0.005 mm., is slightly granulated, and sometimes has a rounded nucleolus. In rare cases, the cell contains some granules along the circumference of the nucleus. These cells are united with each other by their elongated extremities, which sometimes expand like fine ribbons, and at other times, on the contrary, they assume the form of thick cords. A certain distance from the cells these prolongations bifurcate, without being diminished, and a very fine network is formed, and it is uncertain whether it springs from the cells, or whether it is independent of them. These cells vary in size, the smaller being found in the central parts of the retina, and the larger toward the periphery. Their size also varies according to the layer in which they are found. They also vary in configuration, according to the layer in which they exist. In regarding Figures 24 and 25, it is difficult to distinguish that they are of the same character. Their investigation is most easy in the region of the ora serrata, as in that region the cellular tissue largely predominates over the nervous.

The *membrana limitans* is exclusively composed of cellular tissue, and hence it is appropriate to begin a description of the connective tissue from this point. It is a thin membrane, of not more than 0.002 mm. in thickness, except in the region of the ora serrata, where it is thicker. It is a continuous membrane, and is in contact internally with the hyaloid membrane. On

FIG. 24.

Cells of the connective tissue of the human retina. Magnified 300 diameters. (*From Ritter.*)

FIG. 25.

Vitreous metamorphosis of the cells of connective tissue of the human retina. Magnified 300 diameters. (*From Ritter.*)

the outer surface it is rugous, and gives origin to numerous compact fibres, which pass into the retina, and have received the name of *limitary fibres*. This membrane is transparent, and possesses two parallel surfaces, the external of which is interrupted by the fibres. It is distinguished from the vitreous

FIG. 26.

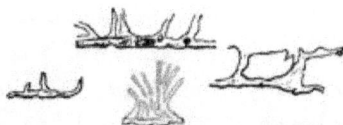

Fibres of the membrana limitans, traced as far as the fibrous layer in the retina of a man. Magnified 300 diameters. (*From Ritter.*)

membrane, to which it adheres, by a peculiar rigidity, by striæ, which extend irregularly from one surface to the other, and by the presence of some disseminated nuclei. The aspect of rigidity which this membrane possesses is caused by the deep coloration of its outlines. Its striæ and its nuclei distinguish its character. These transverse striæ traverse the membrane in various angles, yet they run in pairs, which assume a parallel direction. On the internal surface they are confined to the limitary membrane, whilst externally they are confounded with fibres emanating therefrom, which connect the limitary membrane to the other parts of the cellular tissue of the retina. The nuclei disseminated through this limitary membrane are always found between two parallel striæ. The nuclei of the fibres which are connected with the limitary membrane are not, generally, directly attached to this membrane, but occupy fibres at some distance from it. The size of the cells of the limitary membrane is somewhat uniform, diminishing but little in the central parts of the retina, and enlarge in but small proportion toward the periphery. The length of the cells is, however, variable, being short at the centre and the periphery of the retina, whilst at intermediate points they are found considerably longer. It is but seldom that an entire cell is contained within the thickness of the limitary membrane. The greatest importance is to be attached to its investigation, in consequence of the fibres given off from it. For a long time these fibres have been described as being attached to the limitary membrane by an expanded triangular base, after having commenced by a slender origin, as seen in Fig. 27, from Kölliker; the ex-

Fig. 27.

pansion at *a*, representing the termination of a fibre of Müller on the *membrana limitans*. Ritter is quite certain that this is not correct. By the fibres of the *membrana limitans* is understood only those fibres which can be traced to the fibrous lamina in which they decidedly change in character. These fibres are composed of filiform cells, which communicate with each other by their filiform prolongations. In the central parts of the retina these fibres have a perpendicular course from the *membrana limitans* to the fibrous layer; they are not united with each other, and preserve throughout a breadth uniformly of 0.002 mm., which is identical in thickness with the fibres of the limitary membrane. In the vicinity of the ora serrata, where the nerve-fibres and the ganglion-cells are almost wanting, the size of these fibres increases to two or three times the dimensions above given. They run in various directions, forming various angles, and a network of large meshes. In the most narrow interspaces are found, at intervals, a few nerve-fibres, and in the largest meshes ganglion-cells are found. It is seen that in the middle region of the retina (see Fig. 26) the fibres of the *membrana limitans*, when traced to the nerve-fibre layer, each presents, at its external extremity, the same conical expansion that it has on its internal extremity. Ritter says they are not simple elements, as is indicated by a manifest striation, which indicates complexity of structure.

The size of these fibres depends on the size of the interspaces in the optic nerve layer, and the spaces between the ganglion-cells. The simple fibres are distinct and bright; those in the central parts of the retina generally have small nuclei, difficult to perceive, whilst the larger fibres near the ora serrata contain a great number of large nuclei. Hence, it will be perceived that the fibres of the limitary membrane present numerous varieties of form, partly determined by the proportion of nervous tissue contained in the different parts of the retina. In parts where the latter is less abundant, the cellular tissue seems to take its place. In points where the globular nervous elements exist, as in the ganglion-cell layer, the cellular tissue fills up the cavities, whilst in places where the nervous substance is found in the form of fibres, the cellular tissue forms a network. At the outer limit of the ganglion-cell layer these fibres suddenly expand in all directions, and divide into extremely fine fibres, as seen in Fig. 28, which are so delicate that it is extremely difficult to estimate their dimensions. These fibres form a network in the fibrous layer of extreme tenuity, so that with a moderate illumination it gives to this lamina a granular appearance.

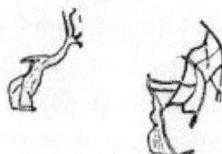

FIG. 28.

Transformation of fibres of the membrana limitans into fibrils of the fibrous layer of the human retina. Magnified 500 diameters. (*From Ritter.*)

The cellular tissue of the fibrous layer presents considerable variations in the different regions of the retina. The fibrous layer is most uniform in its thickness of all the retinal layers. Even at the ora serrata it is this layer that furnishes the larger part of the cellular tissue

found there. In the centre of the retina the fibres are perpendicular to such an extent as to almost deprive the lamina of
its granular aspect, and in the middle of the *macula lutea* this
is most manifest. On the contrary, toward the periphery, the
granular aspect of the retina is more decided, the anastomoses
of the fibres are more marked, and a larger number are seen
running obliquely ; nuclei are also found, which, in appearance,
nearly approach the fibres of the limitary membrane.

The network in the fibrous layer receives the prolongations
of the ganglion-cells, the distribution of which determines the
size of the meshes of this network and their reciprocal disposition. In the centre of the retina the meshes are most narrow,
and the fibres composing it are least oblique, as the prolongations of the ganglion-cells have in this locality a direction
nearly perpendicular to the retinal plane. The meshes of the
cellular tissue increase toward the periphery, in proportion as
the ganglion-cells take, in this region, more volume, and, as
they form numerous anastomoses, the arrangement of the
meshes becomes more complicated. At the external border of
the fibrous lamina the fibres compare with the inter-granular
layer as coarse fibres with distinct outlines. These fibres, notwithstanding their analogy with those of the limitary membrane, are yet distinguished from them, inasmuch as they constantly run in a perpendicular direction to the retinal plane,
and do not form any anastomoses. They also form at certain
points small cavities similar to those found between the cells
of the ganglion-cell layer. These cavities inclose a small
number of cells of the granular layer, and their walls are perforated by the filamentous processes of those cells.

The external prolongation of these cellular fibres form a new
network of fibres, which constitutes the cellular tissue of the
inter-granular lamina. In man the intermediate granular
layer exactly resembles the fibrous lamina, and is distinguished only by perpendicular striæ, which represent the filiform anastomoses situated between the cells and the granules
of the granular lamina. These filiform prolongations belong

to the nervous tissue, and are interlaced by a fibrous network of more narrow meshes than those of the nerve-fibre lamina, as seen in Fig. 29. These fibres assume a regular disposition in forming quadrangular, pentagonal, and hexagonal figures, so as to have the termination and origin of these small elements correspond. The angles of these figures send off anastomoses to neighboring fibres of similar character, so that in the constitution of these figures several fibres are always added. Toward the two surfaces of the inter-granular layer, this network terminates by dark-colored

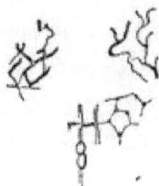

FIG. 29.

Cellular tissue fibres in the inter-granular layer of man. Enlarged 500 diameters. (*From Ritter.*)

fibres, which, in the region of the ora serrata, take the place of the other fibres through the whole thickness of the lamina.

In the granular layer a simple cellular tissue fibre is found proceeding from the fibre-cells, and which are not much finer than the fibres of the limitary membrane. They are distinguished from the fibres of Müller in not inclosing granules, and they contain only a simple nucleus. The enveloping membrane of each cell is largely dilated, and by its external prolongation generally undergoes a vitreous metamorphosis. The reunion of these cells is made by anastomoses in arches. In the central part of the granular layer two series of anastomoses are observed, whilst toward the periphery the whole thickness of the lamina is occupied by a single system of arches, contiguous to the rod layer. As these external ramifications of the cells which constitute the arches are, in great number, the seat of vitreous metamorphosis, in certain preparations the external limit of the granular layer seems to be arrested by a continuous layer. It is this that Schultze named the external limitary membrane. It does not seem proper to name it a membrane, as in the central parts of the retina this expansion of the cells is entirely wanting, and in the middle region it is often interrupted. It can only be termed a membrane in the region of the ora serrata, where it is continuous.

The last fibres of the retinal cellular tissue are situated at the inner termination of the rod and cone layer. They constitute at this point the most external arches, and, as far as is now known with certainty, the cellular fibres of the retina cease there. Ritter thinks it possible that an intimate connection between this cellular tissue and the enveloping membranes of the cones and rods will hereafter be demonstrated.

These investigations of the cellular tissue of the retina were made by Ritter, mainly on the peripheral region, as it is in that region that its characteristics are most conspicuous. Toward the centre of the retina the elements of cellular tissue are very fine, and their direction is parallel to the retinal plane, which readily leads to its confusion with the nervous fibres.

Fig. 30.

External termination of the cellular tissue network of the retina of the whale. Magnified 300 diameters. (From Ritter.)

The retina, as a whole, has a surface of about three hundred square lines. At its peripheral termination at the ora serrata it is firmly attached to the hyaloid membrane. It gradually diminishes in thickness from its central portion to its periphery. At the equator of the eye it yet possesses one-half of the thickness at the centre; but from this point it rapidly diminishes, so that a few lines from the ora serrata it measures only one-third, and at the ora serrata only one-fourth of the central thickness. Ritter says the ora serrata is only a conventional limit, from which point the retina diminishes under an appreciable angle to reflect itself at the distance of two millimetres on the hyaloid membrane as a simple vestige. Up to the equator of the eye all the retinal layers participate equally in the diminution; after this the granular layer and the layer of ganglion-cells disappear. At the distance of four millimetres from the ora serrata it becomes thin, and at the ora serrata not a trace of the nervous tissues remain. At this point the retina is represented only by its cellular tissue, which also is decidedly diminished at this point.

The *macula lutea* is elliptic; has a length of 1'''.44, and a breadth of 0'''.36. With its inner end it is 1.''' to 1'''.2 from the middle of the optic nerve entrance. It loses its yellow color soon after death, and in the fresh eye it can only be seen with the microscope. The layer of ganglion-cells is increased in the macula lutea, eight cells being superimposed on each other. The *rods* are entirely wanting, and are replaced by closely packed *cones*. As regards the granular layer in this spot, the granules are diminished, whilst the layer of granule-cells is increased. The fibrous layer has its normal thickness. The fibres of Müller converge toward the centre, which unusual disposition is explained by the fact that the augmentation in number of the granular cells and of the ganglion-cells shows that the fibres of Müller do not fuse so frequently, as the number of the elements of the *bacillar* layer are not greater than in other parts of the retina. Consequently the elements of the rod and cone layer, which correspond to a ganglion-cell, are less numerous. Hence the fibres of Müller take in the first place a parallel direction, and then a convergent, in proportion as they are nearer the centre of the macula lutea. The increase in number of the granule-cells is so great that the contiguous parts of the retina do not furnish enough of the fibres of Müller, and they are in some measure derived from neighboring parts. The layer of optic nerve fibres is very much diminished, and sometimes imperceptible, as all the bordering fibres end there. Hence, it appears that, anatomically, the distinguishing point between the yellow spot and other parts of the retina consists in the substitution of cones in place of rods. Ritter believes that this particular arrangement of the cones has some relation to binocular vision, which man and the monkey alone possess. The views of Max. Schultze on this point have been referred to, and all things considered, it does seem that he is right in the opinion that the function of the cones is to distinguish color, whilst that of the rods is to furnish quantitative light. The peculiar arrangement of the rods and cones in man, as well as the facts derived from comparative anatomy, favor this theory. Vision is most acute in the yellow spot, where we find cones alone, an increase of granule-cells, and an increase of ganglion-cells, the latter being eight layers in thickness here. Still other parts of the retina also possess vision, being most acute nearest the yellow spot, and gradually diminishing in acuteness to near the ora serrata.

In the macula lutea, a little toward the inner extremity from its centre, there is a colorless, depressed spot of $0'''.08$ to $0'''.1$ in diameter, called the *fovea centralis*. In this depression, according to the measurements of Heinrich Müller and Schultze, the cones are smaller and thinner than in other parts of the yellow spot (where they are smaller than in other parts of the retina), being in the macula lutea $0'''.002$ to $0'''.0024$, and at the fovea centralis only 0.0022 millimetres in breadth. Kölliker says the yellow color is produced by a diffused pigment, saturating the parts of the retina, with the exception of the bacillar layer. The use of the coloring matter in the macula lutea is not known positively. Within the past year Max Schultze (*Ueber den gelben Fleck der Retina, seinen Erufluss auf normales Sehen und auf Farbenblendheit;* Bonn, 1866) published the results of his investigations on this pigment, and he concludes that it is the ends of the rods and cones imbedded in the choroideal pigment that are the perceptive organs of light, and that all perceived light must pass through the yellow coloring matter before it can reach the perceiving rod and cone ends. This yellow-colored point necessarily absorbs some of the violet rays of light of the spectrum. It has been determined that the refracting media have but little to do with the fact that the violet and ultra-violet rays of the spectrum of the human eye can only be seen by a weak light, and hence this feeble illumination must have its cause in the torpid retina; and the yellow coloring matter in the region of most acute vision is intended to bring about the want of irritability of the retina, to modify the bright daylight, so that we do not prefer twilight to daylight, as the owl does, which has scarcely a trace of yellow coloring matter in the retina.

Papilla Nervi Optici.—The optic nerve entrance is round or slightly oval, and has an area of 0.44 square lines, with a diameter of not quite three-fourths of a line.

In the optic nerve itself, the nerve-fibres, with distinct outlines, are united in several fasciculi, among which the thick sheath of the optic nerve sends cellular partitions, which separate them from each other. The largest part of the cellular tissue of the nerve (and especially the outer, thicker layer) is reflected on and blended with the sclerotica. At the inner sclerotical limit, the cribriform plate marks the termination of the cellular tissue of the optic nerve. Donders, however, asserts that the connective tissue envelopes of the nerve-fibres sometimes follow them until within the retina. The *lamina cribrosa*, as its name indicates, forms a sieve. Its meshes are a little serrated, and it is in contact partly with the inner layer of the sclerotica, and with the outer layer of the choroid.

The elements of the cribriform plate consist of cells, identical with those of the stroma of the choroid; sometimes they are intermixed with pigment-cells, which can be recognized by the ophthalmoscope. It is tense and slightly concave anteriorly. In front of the cribriform plate, or, at least, soon after having passed that point, the optic nerve loses its sheath, and also its inner neurilemma, so that the tubules are expanded in every direction, divested of their connective-tissue envelopes.

Seen through the ophthalmoscope, the papilla appears white, like the full moon, surrounded by a red, or dark red, ground, which is the vessels of the choroid. It is sometimes called the blind spot of the retina, from the fact that it is not susceptible to the impressions of light, not possessing the retinal layers essential to the performance of the function of vision, consisting solely of the optic nerve tubules, covered by the membrana limitans.

The central vessels of the retina, the *arteria centralis retinæ*, and the *vena centralis retinæ*, spring from the centre of the optic papilla, as seen in Fig. 31.   About ½''' external to the ring of the optic nerve, outside the eye, the *vena centralis* separates itself from the *arteria centralis*, and takes an oblique, outward direction, whilst the artery continues for some distance a central direction, until it arrives at the optic nerve entrance, within the eye, where it bends in a direction opposite to the vein, in an oblique or knee-shaped manner, to ramify in the retina. In their crossings, sometimes the arteries are in front, and sometimes the veins, more frequently the latter.   Sometimes, in the centre of the papilla, there is a

Fig. 31.

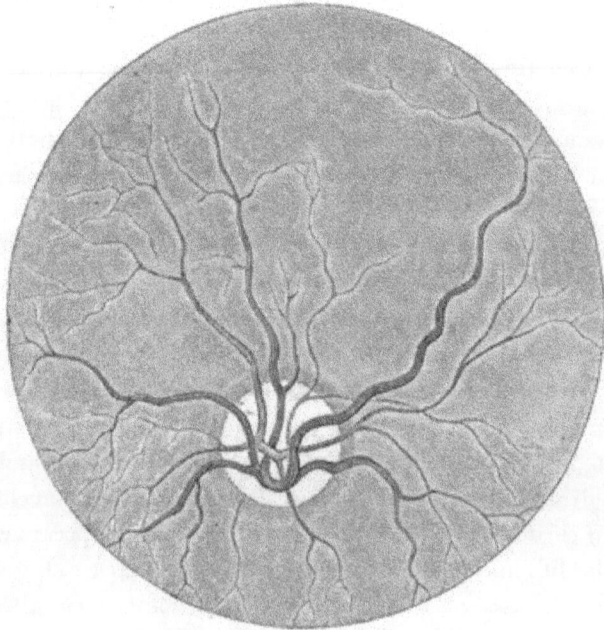

thin covering of nervous matter between the artery and the vein, as seen in Fig. 32, and, as they approach the border of the papilla, they plunge deeper into the nervous matter, and, in their ramification forward, the vessels are completely imbedded by the expansion of the optic fibres.   They continue

to divide into branches as they proceed forward to the ora
serrata, where they pass into a very fine capillary network.
These vessels diminish so rapidly in size, that, at the ora
serrata, no separate vessel can be distinguished. These vessels
send off· no branches into the vitreous body, or choroid, the
retina possessing its own system of vessels. At the beginning

FIG. 32.

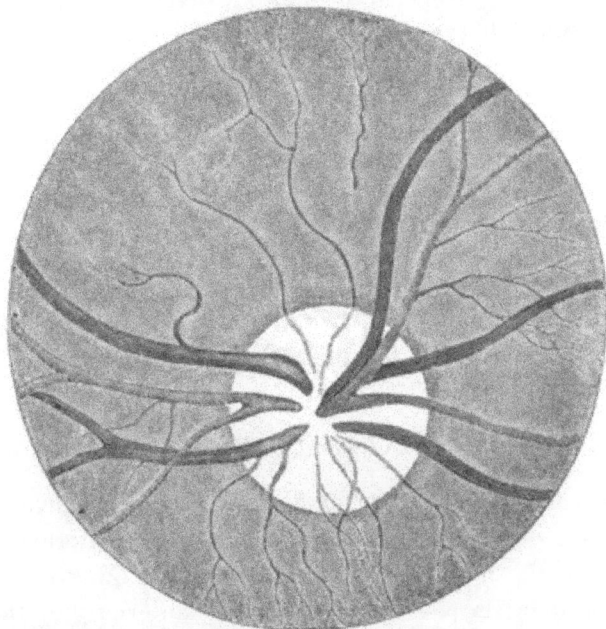

of the *zonula Zinnii* they terminate in a somewhat imperfect
ring of vessels, the *sinus circularis venosus retinæ*, from which
the returning veins proceed. The vessels of the retina have
an intimate connection with the cellular tissue of this membrane. As seen by the ophthalmoscope, the arteries are lighter
and smaller than the veins; the veins are darker, larger, and
more tortuous than the arteries.

The retina still remains a fruitful field for histological and
physiological discoveries. Very much remains to be discovered.
Investigation is needed, and not theorizing. Carl Ritter and

6

Schultze have recently made a real progress in our histological knowledge of the retina. According to Ritter, the impressions of light are first made on the central fibres of the rods and cones, then on the granules, from thence to the granule-cells and to the ganglion-cells, through the optic fibres, to the brain. The rods and cones perceive the most minute luminous impressions, which, in the granules, is converted into a nervous irritation. The granule-cells unite a determined number of the fibres of Müller, and give the sense of color. The ganglion-cells collect all the impressions made on them by a certain number of rods, cones, granules, and granule-cells, and then transmit them to the nerve-fibres, through which the whole is conducted to the brain, where consciousness is enforced. The ganglion-cells, he thinks, may, perhaps, preside over the conceptions of space and form. Still, our present knowledge does not enable us to decide whether the ganglion-cells are endowed with a psychical central action, or whether they possess simply and purely a reflex action.

The Crystalline Lens.

The lens (*lens crystallina*) is like an optical bi-convex lens, and has its only organic connection with the anterior termination of the hyaloid membrane, or *zone of Zinnius*, which sustains it in its place. It is bounded anteriorly by the iris and the aqueous humor, and posteriorly by the vitreous body, resting in a fossa, called the *hyaloid fossa*.

The lens may be considered a rotational body, of which the anterior elliptical segment measures, through its larger axis, $4'''$ to $4'''.1$, and its smaller axis $1\frac{1}{4}'''$ to $2\frac{1}{3}'''$, whilst its posterior surface presents a parabolical curvature of $3\frac{3}{4}'''$ to $5\frac{1}{10}'''$ parameter. It is difficult to give the measurements of the lens, as its diameters constantly vary, as well as the degrees of curvature of the anterior and posterior surfaces, the former varying far more than the latter. The lens is the organ acted on by the accommodative force, and, consequently, its relations to the solid parts surrounding it constantly vary. In the living eye, the distance of the poles of the lens is not constant. The distance from the anterior pole of the lens to the middle of the cornea is 1''' to 1⅓''', and from its posterior pole to the posterior pole of the retina 5⅗''' to 6⅗'''. The weight of the lens is 4 to 4½ grains.

In consequence of the variations of its radial curvatures, as regards other parts of the globe, it is somewhat difficult to determine accurately its relative location. Pilz lays down the following rules:

A line drawn through the anterior pole of the lens, perpendicular to the eye-axis, touches the insertion of the *tensor choroideæ*, and is 1''' distant from the centre of the cornea.

A line drawn through the posterior pole of the lens strikes the origin from the choroid of the *corpus ciliare*, and measures 8''', and at its centre is 3½''' distant from the middle of the cornea, and 6⅗''' to 7⅗''' from the posterior pole of the eye.

Two-thirds of the *processus ciliares* fall anterior to a line drawn through the equator, perpendicular to the axis of the eye.

The iris rests on the lens, and its degree of anterior curvature is dependent on the extent to which the lens projects.

Helmholtz says that the axis of the lens does not correspond with the anterior pole of the cornea, as the centre of curvature of the latter is located to the nasal side of the lens-axis.

The lens is a perfectly transparent body, and is to be distinguished into the *capsule* and the *lens proper*.

The capsule of the lens (*capsula lentis*) is a perfectly transparent, structureless membrane, elastic, yet easily torn, and surrounds the lens like a mould, completely inclosing it on all sides. The capsule of the lens measures, at its anterior wall, 0'''.005 to 0'''.008, which thickness does not extend as far back as the equator of the lens, but immediately behind the attachment of the *zonula Zinnii* it becomes thinner, and measures only 0'''.002 to 0'''.003.

The thickness of the lens-capsule varies very much, and

seems to increase with old age. It is always thinnest at the posterior pole, in the adult measuring at that point, according to Becker, 0.009 mm.

Although the lens-capsule is a continuous membrane, yet, for the sake of convenience, it is described as consisting of an anterior and a posterior capsule. Histologists do not agree on the structure of the capsule. Kölliker, Pilz, and Becker assert that it is striated, which indicates that it is lamellated. Becker says that this striated appearance is seen in the larger ruminants. It was seen by Leydig, in the heifer, and by Kölliker, in man. Becker asserts that, when sufficiently magnified, this superficial striation is always seen. On the contrary, Ritter, Stellwag, and Hulke confidently assert that the capsule is wholly structureless. The latter writer says that the delicate converging lines which stretch from the attachment of the suspensory ligament, for some distance toward the anterior pole, and the similar, but fainter, lines upon the outer surface of the posterior half of the capsule, are due to the peculiar arrangement of the fibrous cordage of the suspensory ligament. When torn, the capsule contracts by its own elasticity, and curls in, so as sometimes to permit the lens to escape spontaneously.

FIG. 33.

Section of the capsule of the lens of a calf, at the locality where the embryonic cells are situated. Magnified 230 times.

(From Becker.)

The posterior surface of the anterior capsule is lined by a delicate layer of pavement epithelium, which is located between the capsule and the lens proper.

Becker says that, strictly speaking, it is not correct to say that the posterior wall of the anterior capsule is lined by an epithelial membrane, inasmuch as beneath the attachment of the zonula are found irregular granules of various sizes, and of clearly defined outlines, closely packed, and which have around them but a small quantity of *protoplasma*. They often possess distinct divisions. Everything points to the fact that the bodies in question are quite young, and, to a certain extent, imperfect cells, which are destined for a further metamorphosis, and hence are named embryonic cells (*bildungs zellen*). Toward the anterior part of the capsule, these cells gradually are changed into a pavement epithelial covering; and backward, toward the equator of the lens, are found the small round cells, which, still further back, sprout out into true lenticular fibres, as will be explained hereafter.

These nucleated epithelial cells also perform an important office in connection with the nutrition of the lens.

Immediately opposite, on the outer side of the capsule, are placed the *processus ciliares*, which are made up in structure, almost wholly, of vascular loops; and osmotic circulation will readily take place through so permeable a membrane as is the lens-capsule.

Hulke, who seems to have very thoroughly investigated the human lens, agrees with Becker, and, for convenience of description, divides the anterior hemisphere of the capsule into a central region, immediately around the anterior pole and the marginal region, extending outward as far as the edge of the lens. In the central region, the epithelium is a delicate pavement, formed by a single layer of large polygonal cells, joined

Fig. 34.

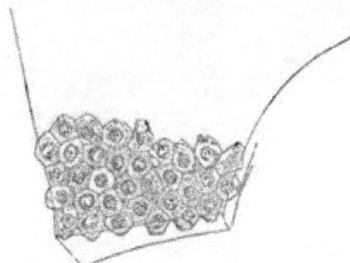

Intra-capsular epithelium, central region. (*From Hulke.*)

edge to edge; the cells are conspicuous for their sharp outlines, and each contains a large circular nucleus, which, in turn, incloses two or three small dark dots or nucleoli. The nucleoli are remarkable for their uniform size and regular circular outlines; they occupy the centre of the cells, and each nucleus is separated from the neighboring ones by a space about equal to its own diameter.

In the marginal region the epithelial cells are much smaller and more closely packed together. The nuclei are smaller, have a less regular outline, and are separated by extremely minute intervals; indeed they are sometimes almost in contact, and the walls of their containing cells can hardly be made out. The epithelium in this region is not arranged as a simple pavement, but rather in the form of a bed, in which the cells are crowded in superimposed layers; the bed is not prolonged beyond the edge of the lens, and the inner surface of the posterior hemisphere.of the capsule is void of epithelial lining.

The transition from the large polygonal cells of the central region to the small crowded ones of the margin is not abrupt; every possible gradation occurs between these extreme forms.

Fig. 35.

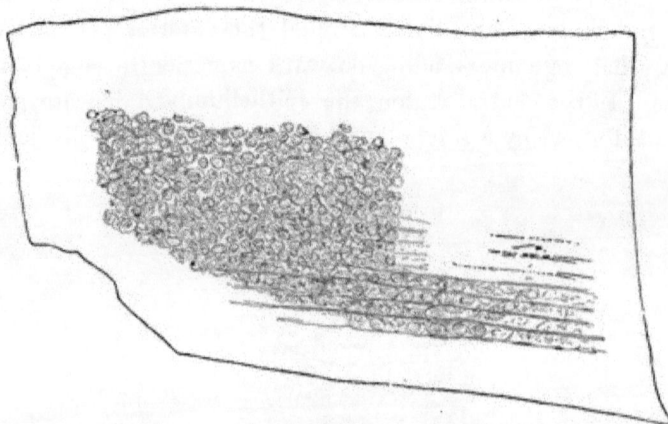

Intra-capsular epithelium of the marginal region, constituting the matrix of the lens.
(*From Hulke.*)

This epithelial bed at the margin of the anterior half of the capsule is the matrix of the lens; in the young it contains free

nuclei; and it is the growing part in which the lenticular fibres
are formed by the growth and metamorphosis of the cells.

Becker says that the smallest of these embryonic cells
are found in groups in the *protoplasma*, from two to six in a

FIG. 36.

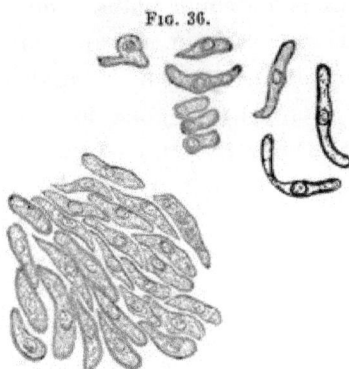

Cells of the matrix undergoing transformation into lenticular fibres. (*From Hulke.*)

group, as seen in Fig. 37 at *a*, which are constantly being
crowded back.

FIG. 37.

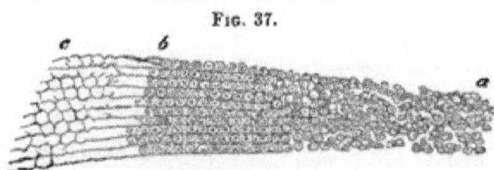

Their nuclei become larger, more round, isolate themselves
more, and take such a position, that, finally, they are found
in rows behind each other, as seen at *c*, Fig. 37.

Up to this time there are no well-defined cell-walls, but
from this point they are recognized. These cells, arranged in

FIG. 38.

Vertical lens-fibres seen from the capsule. From a dove's eye. Magnified 230 times.
(*From Becker.*)

rows, begin to grow in length, and still further separate themselves. After having reached a certain length they form neat

convolutions, as they assume a more oblique direction backward. These fibre-cells, or lens-fibres, thus originated, are
crowded more and more inward to the lens, so that finally
their position is changed, and they enter into the concentric
laminæ of the lens as lenticular fibres. (See Fig. 36.)

The embryonic cells, which, at first, are round, as they
increase in volume, being surrounded on all sides by growing
cells, necessarily assume a hexagonal form, as seen in Fig. 38.

At the point where they curve, the strongest pressure
naturally takes place in front, which, therefore, is more thin
here, whilst back, toward the capsule, they expand more.
The further inward, toward the centre of the lens, that the
fibres are situated, the more flattened they become, until
they are found as flattened hexagonal bands. Their ends,
anteriorly as well as posteriorly, overlap each other in the
form of shingles on a roof, as seen in Fig. 39.

FIG. 39.

Ends of lenticular fibres abutting on the lens-capsule.   Side view.   From the lens of
a calf.   Magnified 230 times.   (From Becker.)

In the neighborhood of the equator of the lens the ends of
the fibres are broader than the fibres themselves, as seen in

FIG. 41.

FIG. 40.

Showing serrated edges of the
fibres of the lens. (Becker.)

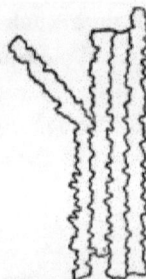

Serrated lenticular fibres
from the lens of a cod.

Fig. 39. Further back the fibres become more thin and narrow, and on the edges are usually serrated, as seen in Fig. 40.

In the fish and amphibia the fibres are much more serrated, as seen in Fig. 41.

As soon as the lenticular fibres have reached the star, their
natural boundary, that prevents further growth, a change
takes place in their form. Instead of continuing their course,
the ends turn nearly perpendicularly away from their former
course, as seen in Fig. 42, Fig. 43, and Fig. 44.

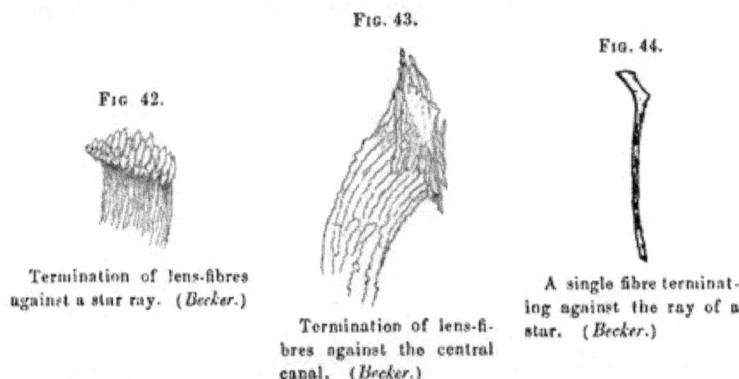

FIG. 43.

FIG. 44.

FIG 42.

Termination of lens-fibres
against a star ray. (*Becker.*)

Termination of lens-fi-
bres against the central
canal. (*Becker.*)

A single fibre terminat-
ing against the ray of a
star. (*Becker.*)

In Fig. 44 there is a single fibre showing the manner in
which the ends curve. The walls of the star are pressed
closely together; still the hexagonal form of the ends remain
visible.

The youngest fibres have but a very delicate enveloping
membrane, which is easily ruptured, and allows its albuminoid contents to escape. Later, the contents of the fibres
become more solid, so that it will escape only from its ends.

In the older text-books there is given a description of the
liquor Morgagni, a thin albuminoid substance between the
lens and the capsule. More recently it has been discovered
that this liquid is collected by some post-mortem change,
supposed to be a solution of the epithelial cells.

Becker says that the epithelial cells are not readily dis-
solved, and he believes the *liquor Morgagni* to be composed
of the escaped contents from the ends of the lenticular fibres,
along with the substance contained in the stars.

At birth the multiplication of new fibres is quite active, but diminishes as age advances; the zone of embryonic cells becomes more narrow, yet the new formation of lenticular fibres progresses, though slowly,—the deep-seated fibres near the centre of the lens atrophy, the lens becomes more solid, less elastic, and more flattened; all of which, obviously, has its influence on the accommodative act, with such regularity that it may be denominated a fixed physiological law, as Donders has taught that emmetropic eyes have a certain accommodative power at a certain age. The matrix being situated at the equator (a constant deposition of new fibres taking place there) accounts for the flattening of the lens in old age. The *lens* itself is composed of the lenticular fibres and the interstellar and the interfibrous substance (Becker). The lenticular fibres are flat, six-sided elements, which, according to Kölliker, have a breadth of $0'''.0025$ to $0'''.005$, and a thickness of $0'''.0009$ to $0'''.0014$, and are perfectly clear, pliable, and soft.

Their breadth and thickness is modified by age, and by the position of the fibres, whether located near the nucleus or near the surface. Near the nucleus the fibres are mere flattened bands in adult persons. (See Fig. 45.)

FIG. 45.

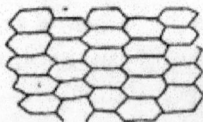

Layer of lenticular fibres cut vertically so as to show their hexagonal form.

They are delicate walled tubes, containing a clear, viscid, albuminoid matter, and they become darker and more distinct in all substances that coagulate albumen; and therefore chromic acid, nitric acid, alcohol, and sulphuric acid, are used to harden the lenticular matter to facilitate its investigation.

For a long time, the lens has been divided by anatomists into a cortical portion, and the nucleus. According to recent histological investigations it is not strictly correct. It is true that in a certain sense we have a nucleus, and a cortical portion; but there is nothing like a distinct separation between the two, but a gradual change from the soft tubular condition of the surface, to the more solid, fibrous structure of its central part.

The lens has also been described as lamellated,—consisting of layers like an onion. This is true only to this extent;—the lenticular fibres separate more readily on their external or internal surfaces than on their sides, where the surface is serrated. So there is nothing like regular layers, but any number of fibres may be peeled off, according to the character of the force applied. In short the lenticular fibres constitute the fundamental elements of this organ, being composed of the superinduced fibres, from the very centre to the periphery.

The *stellate figures* have quite different forms at different periods of life. We find them in their most simple form in the fœtus, and in new-born children, with a star of three rays, which regularly meet, on the anterior surface of the lens, in an angle of 120°, with two rays pointing obliquely downward, and one vertically upwards. On the posterior surface of the lens this figure is inverted, the two rays being directed obliquely upward, and the vertical ray, downward, in the form of the letter *Y*.

FIG. 46.

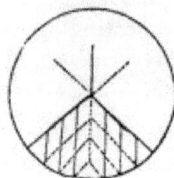

In Fig. 46, taken from Pilz, the dark lines represent the anterior surface, and the dotted lines the posterior surface.

The posterior star, compared with the anterior, appears as if turned round through an angle of 60°. The direction of the lenticular fibres in the individual laminæ is as follows: No individual fibre traverses the entire semi-circumference of the lamella, as a fibre starting from the axis of the lens, either on its anterior or posterior surface, will not reach the lens-axis of the opposite surface, but immediately after curving round the equator, will attach itself to the end of the ray close to the equator. Or, reversing it, say a fibre starts from the very end of a star-ray, it will curve round the equator and terminate at the lens-axis of the opposite surface.

FIG. 47.   FIG. 48.

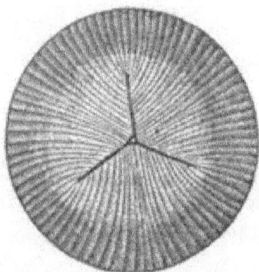 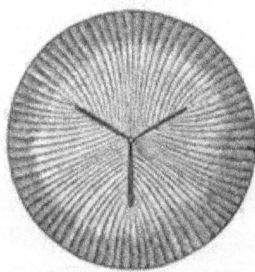

Fig. 47 represents the anterior surface of the lens of a child, and Fig. 48 the posterior surface. (*From Nunnely.*)

On looking again at Fig. 46, it will be observed that the longest dark lines representing the fibres on the anterior surface, are the shortest on the posterior surface, represented by the dotted lines, and *vice versa*. This is the mode of extension of all the lens-tubes, none of them going quite round, and all those which lie in one layer being of equal length. In the adult the nucleus of the lens presents exactly the same condition, but on the other hand, in the superficial lamellæ, and on the surface itself, a more compound star is found, having from nine to sixteen rays, of various lengths, and rarely quite regular; still, however, certain main rays may even here be distinguished from the others. (See Fig. 49.) The course of the fibres necessarily becomes more complicated by this means; and the more so, as on such stars the fibres attached to the side of the rays

converge in an arcuate manner, giving rise to a penniform or
whorled appearance (*vortices lentis*). But nevertheless, the essential points of the course of the fibres just described remain
completely the same, seeing that in these more complex stars
the rays of the anterior and posterior aspects do not correspond, and that no fibre goes from one pole to the other.

Fig. 49.

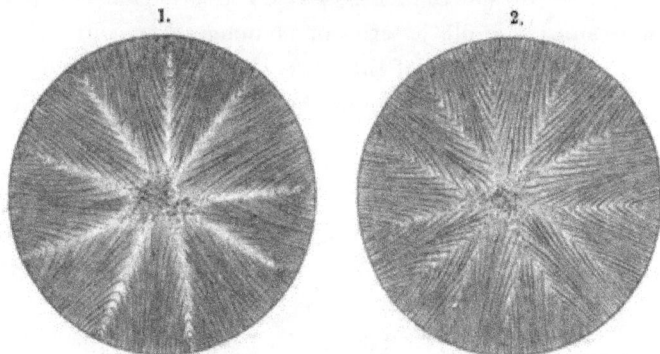

Lens of the adult, after Arnold, to show the stars. 1. Anterior surface ; 2. Posterior
surface.

As the stars pass through all the laminæ, there exist three
or more perpendicular non-fibrillated planes, called central
planes by Bowman (Kölliker). The manner in which the
fibres terminate at the stellate figures, has been alluded to
above.

Fig. 50, from Becker, shows their mode of termination.

Fig. 50.

Terminations of lenticular fibres, the large ends terminating in the stars. (*Becker.*)

In the stars, says Kölliker, the substance of the lens is not
formed of tubes, as elsewhere, but consists of a material, which
is, in part, finely granular, in part homogeneous.

Becker, who seems, most patiently, to have looked into this

matter, says, that in the fresh state, the substance in the stars is thick, altogether homogeneous, as clear as water, and of the same index of refraction as the fibrous part. It becomes finely granulated by coagulation, by boiling, also, in the mineral acids, and in alcohol, and in alkalies is re-dissolved, and has the general properties of protein substances.

This substance is not confined to the stars, but is wedged in between the fibres of the lens. Becker asserts that this homogeneous substance fills a series of channels that communicate in the equatorial region of the lens.

Fig. 51 shows the interfibrous channels in the lens of the calf.

FIG. 51.

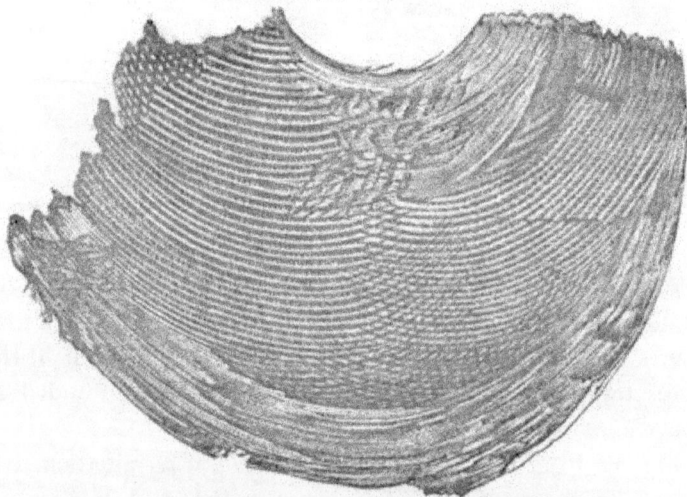

Radiary splitting of a lamella of the lens, embracing nearly the half of it. It shows the interfibrous channels, which are seen to cross each other. Magnified 65 diameters. (*From Becker.*)

The homogeneous substance filling the interfibrous channels is coagulated, and is represented by the dark lines.

Corpus Vitreum.

The histological structure of the vitreous body is but imperfectly understood, and it has for a long period been the object

of animated discussions among microscopists. Its structure
seems to be a peculiar one in the economy, and in the attempt
to place it among the various tissues it resembles, authors have
most widely differed.

The descriptions of Hanover and of Finkbeiner seem to be
sustained better, by more recent histologists, than those of any
other authors. Their views are not adopted, however, by some
high authorities. Further investigation must be awaited,
before anything like a positive histological description can be
given of this body.

The vitreous body fills up the space between the lens and
the retina, and through the *zonula Zinnii* is connected with
the lens. It lies loosely on the retina, except at the optic
nerve entrance, where its connection is more intimate, and
with the *corona ciliaris* and the lens it is quite firmly connected.

The enveloping membrane (*membrana hyaloidea*) is an extremely delicate membrane, scarcely perceptible under the
microscope, and measures 0'''.002. This is true, only of that
portion back of the *ora serrata;* in front of this point, it becomes more firm, and is known as the *pars ciliaris hyaloideæ
seu zonula Zinnii*, and named by Retzius, *ligamentum suspensorium lentis*, and proceeds to the border of the lens, to become
blended with its capsule, in the manner hereafter to be described.

In structure, the vitreous body is a clear glass-like (*vitrina
ocularis*) substance, described by Virchow as a homogeneous,
muciferous substance, and by Kölliker as belonging to the
primitive forms of gelatinous connective tissue, bearing a considerable resemblance to the enamel organ of the embryonic
dental sac. This substance is enveloped in a system of membranes, which proceed from the surface of the hyaloid membrane, and toward the centre, in the form of radii, dividing
the vitreous humor into sections like an orange.

Hanover made this discovery, on eyes immersed in chromic
acid, and later, Finkbeiner corroborated this, by investigations
made on eyes immersed in a solution of bichloride of mercury.

All the sections converge toward the optic axis, which is the space occupied in the embryo by the *arteria centralis* in the *canalis hyaloidea*.

This *canalis hyaloideus seu Cloquetii* has its beginning in the region of the *papilla nervi optici*, and is named the *area Mortegiana*, and is the common axis of all the sections. The sections do not quite reach the axis of the eye, but a cylindrical space is left centrally, without membranous structure, and which is considerably larger in the child than in the adult.

The walls of the sections gradually become more delicate, until they can no longer be discerned. Hanover counted 180 radii in the human eye, and he believes that to be the sum of the sections composing the vitreous body.

Finkbeiner claims that the section walls are the same in structure as the membrana hyaloidea, having on each side a layer of pavement-cells.

This division into sections, as taught by Hanover, and later, by Finkbeiner, is sustained by Pilz, Bowman, Heiberg, and others.

Pilz says this cannot be an error, such as Brücke was led into, when he thought the vitreous body laminated, resulting, Pilz thinks, from the effect of the solution of the acetate of lead, which he used as a hardening remedy. The hyaloid membrane must be divided into the portion that envelops the vitreous body, and that portion extending from the beginning of the ora serrata to the capsule of the lens.

Whilst this membrane, in the back part of the eye, envelops the *corpus vitreum* as a single membrane, it divides itself, in the region of the *ora serrata retinæ*, into two plates, of which the posterior one forms the anterior wall of the vitreous body, its anterior surface forming a dish-shaped concavity (*fossa hyaloidea*) for the reception of the posterior surface of the lens; whilst its posterior surface forms the basis for the sectional divisions above described.

The anterior division proceeds forward, toward the lens, its outer surface being covered by the ciliary processes, and before it reaches the equator of the lens, it again divides into two laminæ,—an anterior and a posterior,—the former being corrugated and proceeding to the anterior lens-capsule, with which it becomes blended; the latter going to the posterior capsule, uniting with it in the same manner. The anterior corrugated, plaited lamina, on which rest the processus ciliares, is the *zonula Zinnii*. The triangular space around the lens equator, formed by the last-named division, is *Petit's canal*, or, the *canalis godrome* (Fig. 1, C P). In this division, it will be observed that between the anterior surface of the posterior lamina, forming the *fossa hyaloidea*, and the posterior surface of the posterior lamina of the second division, going to the posterior lens-capsule, another longer, but more narrow canal is formed,—the *canalis Hanoveri* (Fig. 1, C H). According to Finkbeiner, the hyaloid membrane is composed of a fibrous texture, covered with an epithelial layer.

The former is composed of an innumerable mass of delicate elementary connective tissue fibres tied together, which have a fine striation, and finally, in their course, end in true connective tissue. These fibres are too fine for measurement, and in acetic acid they swell, and finally disappear, leaving finer, darker and shorter threads, which are supposed to be nuclei.

At the ora serrata, where the membrana hyaloidea is intimately connected with the retina, these fibres become more visible, unite, and form a texture, very similar to connective tissue. Further forward, beneath the processus ciliares, they unite more intimately, and anastomose, until broad bundles are formed, which from hence proceed to the lens-capsule in two forms, which there expand as bands and fibres, to unite with the capsule; or beneath the ciliary process, they divide dichotomously, and are united to the lens-capsule by fine fibres, which can be traced from the ciliary processes.

Beneath the processus ciliares, Finkbeiner discovered a transverse striation of these fibres, which Retzius had already discovered. Toward the borders of the ciliary processes it begins

7

to fade, and in the thinnest part of the zonula Zinnii it cannot
be discerned. Finkbeiner could not demonstrate that they
are muscular fibres. Pappenheim, Ammon, Henby, Sappey,
Bendz, Frey, Kölliker and Weber agree that these fibres resemble connective tissue, or elastic fibres. Heiberg, who has
recently gone over this ground, concludes that the zonula does
not suddenly divide at the ora serrata, but that even back of
that point, fine fibres originate, which, by lateral anastomoses,
proceed forward, to form beneath the ciliary processes more
firm bundles, which finally unite in a continuous membrane,
which follows all the elevations and depressions of the processus
ciliares, and at the same time connecting with the parts beneath,
so that the pars ciliares retinæ and the choroidal epithelium all
fall together when the zonula is forcibly detached. Just before reaching the ciliary processes, this membrane divides,—
one plate going to the anterior capsule of the lens, and the
other to the posterior capsule, thus forming between them, the
canalis Petiti.

This would still leave the true hyaloid membrane, which
proceeds forward, and forms the anterior wall of the vitreous
body and the hyaloid fossa. Between the posterior lamina of
the *ligamentum suspensorium lentis* and the hyaloid membrane,
the *canalis Hanoveri,* described above, is formed.

When the parts are removed, so that the connection between the zone of Zinnius and the lens can be seen, it will be
perceived that it surrounds the lens like a belt. The fibres at
the lens-capsule expand, and in a zigzag form are connected
with it. Each bundle of fibres with its pointed surface is connected with a ciliary process. The zone has, as a connective
material, a thin, structureless, vitreous membrane, much corrugated, with two kinds of fibres,—those with longitudinal
striæ, and the more numerous, with transverse striæ. (See
Fig. 52.) The fimbriated end is attached to the capsule.

The fibres seem mostly to originate from the membrana
hyaloidæ, but a portion of them seem connected with the
cells of the pars ciliares retinæ, which Kölliker asserts to

be connective tissue, and which is a continuation of the radiary or connective tissue fibres of the retina. Heiberg has not seen

FIG. 52.

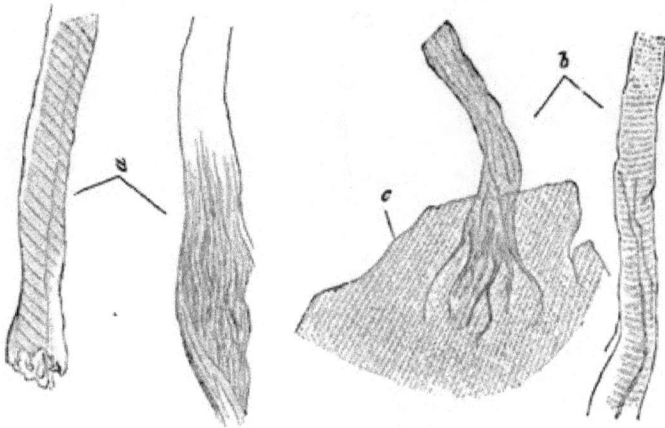

Fibres from the *zonula* of man; *a*, prepared in Muller's solution; *b*, prepared in solution of nitrate of silver; *c*, a piece of the capsule. Magnified 300 diameters. (*From Heiberg.*)

these cylindrical cells in man, in the process of transformation into the zonular fibres, but believes he has discovered them in the sea-dog. In separating these cells, their inner ends are often seen with anastomosing points, as in Fig. 53, and sometimes they terminate in a thread-like process; as in Fig. 54.

FIG. 53.

Cells from the *pars ciliares retinæ*, the points, *a*, anastomosing. Human Eye, magnified 350 diameters. (*From Heiberg.*)

Heiberg could not positively determine in man, that the fibres of the zonula Zinnii are muscular, his only test being the microscope. In the horse, he is satisfied that he discovered

muscular fibres in the zone; and analogy, as well as the results
of his microscopic investigations, cause him to believe that

Fig. 54.

Showing a thread-like process, *b*.   (*Heiberg*.)

there are muscular fibres also in the zonula of man. It has
contractile tissue. It is one of the important parts concerned
in the accommodative process. The contraction of the zonular
fibres in connection with the radiary fibres of the ciliary muscle, extends, or enlarges the equatorial diameter of the lens,
and diminishes the antero-posterior diameter, and fixes the
eye for remote vision. These two sets of fibres are the active
agency in the so-called negative accommodation. They are
doubtless antagonistic to the circular fibres of the ciliary
muscle.

When the circular fibres of the ciliary muscle contract, they
overcome the tension of the zonular fibres, and the lens, by
its inherent elasticity, has its antero-posterior diameter enlarged, and the eye is fixed for near vision.

The circular fibres of the ciliary muscle, then, are the active
agency concerned in vision for near objects, or in positive accommodation; the fibres of the zone of Zinnius in connection
with the longitudinal fibres of the ciliary muscle, are the active
agency in accommodation for remote objects, or for negative
accommodation. This agrees with the observations made by
Otto Becker of Vienna,—that during the accommodative act,
the ciliary processes retract, and during life, at no time, do
they proceed as far forward as the lens equator, and that by

their contraction they cannot cause pressure on the equatorial surface of the lens, and thus cause the sides to bulge out, increasing the antero-posterior diameter of the lens, and fixing the eye for the near point of vision, as has been the favorite theory of many ophthalmologists. In its physical characteristics, the vitreous body is colorless, perfectly transparent, highly elastic, yielding, but not very compressible. By filtering, it separates into a fine hyaline substance, and into a fluid clear as water, thin and slimy. In form, the corpus vitreum presents a not quite regular rotational ellipsoid, of which the larger axis in the larger diagonal diameter of the ball is $9\frac{2}{3}'''$ to $10\frac{1}{4}'''$, the vertical axis in the diameter of the bulbus oculi is $9\frac{1}{2}'''$ to $9\frac{2}{3}'''$. The depression in its anterior wall, the fossa hyaloidea, is $4'''$ in diameter, and is quite round. The optic axis through the vitreous body, is shortened $0'''.2$ to $0'''.3$ by the hyaloid fossa. The connections of the vitreous body with the surrounding parts are at the entrance of the optic nerve, at the ora serrata retinæ, where it is connected with the retina and choroid, and, through the zone of Zinnius, it is connected with the lens. It never contains as much fluid as its volume seems to permit, and hence it allows the alterations in size of the lens diameter during accommodation, and the changes in form that must follow the action of the muscles; and its inherent elasticity enables it to resume its form.

Hanover says, that the division of the zonula, going to the posterior lens-capsule, is not corrugated. This seems to be a mistake.

Heiberg says, that many of the fibres of the zone of Zinnius, when they arrive at the point of separation to send a plate to each lens-capsule, divide dichotomously, one branch going to the anterior surface, and the other to the posterior, of the lens, to be there expanded and attached. On the posterior surface they are, however, distinctly seen only on the periphery of the capsule. That there is not an intimate union between the posterior laminæ of the zone and the hyaloid membrane, forming the hyaloid fossa, is obvious, from the fact that the cataractous lens is sometimes removed, with its capsule, leaving its bed in the hyaloid fossa with an unruptured membrana hyaloidea.

The vitreous body in the adult is wholly without bloodvessels or nerves. In the embryonic state it has a system of vessels of its own, which at or before birth rapidly shrink or disappear. In the fœtus it is also richly supplied with cells, which after birth rapidly disappear, until few are left. They are partly composed of round or oval, finely granular nucleated cells, of cystoid character, and partly of larger cells, with several nuclei, with well-defined boundaries. They are found particularly near the ora serrata, behind the lens, and in front of the optic nerve entrance. According to Bowman and Iwanoff there is a fine dense net-work of fibres with dark nuclear corpuscles at the points of interlacement. This view is being corroborated by Stellwag and Kölliker.

In the adult the nutritive plasma is, according to Pilz, derived exclusively from the bloodvessels of the ciliary processes.

The Aqueous Humor.

The Aqueous Humor (*Humor Aqueus*) fills up the space between the *zonula Zinnii*, the *processus ciliares*, the anterior capsule of the lens, and the cornea, and weighs from $3\frac{1}{2}$ to 5 grains, with a specific gravity of 1.0053. It is thin, clear, slightly viscid, colorless, without smell, with a slightly salty taste. It does not distend the chambers to a high degree, so as to permit the necessary excursions of the surface of the lens during accommodation, and the movements of the iris.

It consists chemically of 98.687 parts water, and of solid matter 1.313 (0.467 organic and 0.486 inorganic), of which there are albuminates 0.1223, extractive matter 0.421, table-salt 0.689, chlorate of potash 0.0113, sulphate of potash 0.0221, earthy phosphates 0.0214, and lime 0.0259.

The iris divides the cavity occupied by the aqueous humor into two chambers, the larger called the anterior chamber, and the smaller, the posterior chamber of the eye. The posterior chamber is the space between the ciliary muscle, ciliary processes, and the anterior surface of the lens, and the *uvea*, or posterior surface of the iris.

The anterior chamber is the space between the anterior surface of the iris and the cornea. The chambers communicate through the circular opening in the iris, the pupil. According to Budge, the posterior chamber has $\frac{1}{4}$ the diameter of the anterior chamber. Petit gives as the distance from the axis of the pupil to the lens $\frac{1}{2}'''$ to $\frac{1}{4}'''$, at the periphery of the pupil $\frac{1}{4}'''$ to $\frac{3}{8}'''$, and in the anterior chamber from the axis of the pupil to the middle of the cornea $\frac{1}{2}'''$ to $1'''$. Whilst it has been denied by Stellwag, Cramer, and others, that there is a posterior chamber, it cannot be denied that a small portion of aqueous exists between the lens-capsule and iris; in the frozen eye a small pellicle of ice will be found between the two. However, the volume of the chambers constantly varies, according to the position of the lens and the curvature of its surface, as is evident on observing the variations of size in the anterior chamber during fixation for the far and near points of vision, which also proves a laxity of tension in the chambers permitting these variations.

The quantity of aqueous humor between the iris and the capsule of the lens is smallest in the narrow pupil; and in the dilated pupil this distance is considerably increased.

The opinion that formerly prevailed, that the aqueous humor is secreted from a serous membrane lining the chambers, is now known to be an error. It is derived from the vessels of the ciliary processes, and not from those of the iris. When lost, it is re-accumulated in a very brief time.

In the operation of *paracentesis oculi*, the surgeon frequently empties the chambers at intervals of from five to ten minutes. It is not known whether it is continually secreted and absorbed, or whether such speedy regeneration ensues only after a loss thereof.

The Orbit.

The eyeball lies in the bony cavity, the *orbita*, surrounded by a soft cushion of adipose tissue, and is sustained by the muscles of the eye, and by the processes of the tunica vaginalis bulbi. Through the optic nerve, which enters the orbit through the foramen opticum, and through sensitive, motor, and ganglionic nerves, it is connected with the brain and spinal marrow, and through the conjunctiva, with the lachrymal apparatus and the lids.

The orbits, the bony sockets in which the eyes are lodged, are two pyramidal cavities of irregular quadrilateral form, with their bases directed forward and outward, and their apices backward and inward, so that their prolongation backward would form an angle of 45° on the *sella Turcica*. The temporal side of the base does not project so far forward as does the nasal side, which permits a wider lateral field of vision. The superior boundary or roof is arched, and is formed by the orbital plate of the frontal bone, and by a part of the lesser wing of the sphenoid bone, the base of which is traversed by the optic foramen, a little to the inner side of the apex of the orbit, and gives entrance to the optic nerve and the ophthalmic artery. There is to be noticed on this upper boundary, the roughness or spicula for the attachment of the pulley of the superior oblique muscle, situated at its inner and anterior part; also the depression for the lodgment of the lachrymal gland, at the outer margin of the angular process of the frontal bone, just within the margin of the orbit; also the supra-orbital notch, or foramen, for the exit of the supra-orbital artery and the frontal branch of the trigeminus nerve. The inferior boundary of the orbit is formed by a part of the malar bone, and by the orbital processes of the superior maxillary and the palate bone, and is inclined downward and outward, and forms the roof of the *antrum Highmori*. There is to be observed a groove, which runs forward and ends in the *infra-orbital foramen*, which gives passage to the infra-orbital branch of the internal maxillary artery, and to the middle and anterior dental nerves, being branches of the maxillary division of the trigeminus.

The *internal* boundary of the orbit is formed by the lachrymal bone, the os planum of the ethmoid bone, and part of the body of the sphenoid bone. There is to be observed in this boundary, immediately behind the border of the *processus frontalis* of the superior maxillary bone, the *fossa lachrymalis*, which terminates below in the *canalis nasa lachrymalis*, which is formed posteriorly by the ungual bone, and where it terminates in the inferior meatus of the nose, it is completed by the inferior spongy bone. There are also to be noticed here the anterior and posterior ethmoidal foramina, for the passage through the first of the nasal nerve and the anterior ethmoidal artery, and through the latter, the posterior ethmoidal artery and vein. The external boundary of the orbit is formed by the orbital process of the malar bone, and the great ala of the sphenoid bone. There are here some small foramina, through which junctions of the branches of the fifth and seventh nerves are made. We have further to observe at the superior external angle of the *fissura orbitalis superior*, and at the outer inferior angle, the longer but more narrow *fissura orbitalis inferior*, through the former of which enter the third, fourth, first divisions of the fifth, and sixth nerves, and through it the ophthalmic vein passes out of the orbit. The latter fissure is filled with fat, and gives passage to the *infra-orbital* nerve and bloodvessels. The orbit is lined by a thin and rather loosely connected bone-membrane, the *periorbita*, and it is in connection with the *dura mater* through the *foramen opticum* and the *fissura orbitalis superior*, and with the periosteum of the bones of the face, through the *fissura orbitalis inferior*, the *canalis zygomaticus facialis, foramen supra-orbital* and *canalis infra-orbitalis*. This *periorbita* sends out fibrous processes, partly to the *tunica vaginalis bulbi*, and also to the *ligamentum palpebrale internum*, and *externum*, and is here connected with a mass of firm connective tissue containing but little fat; it also sends out processes to the lachrymal sac, to the lachrymal gland, to the tendon of the *levator palpebræ superioris*, and the tendon of the *obliquus superiori*; at the border of the orbit it passes over to the outer convex edge of the cartilage of the lid, the *ligamentum tarsi latum*, coming here in connection with a process of the *tunica vaginalis bulbi*, which will soon be described.

At the optic foramen, the *periorbita* is condensed into a thick fibrous ring, the *annulus fibrosus*, which forms an elliptical ring around the *foramen opticum*, and the middle part of the *fissura orbitalis superior*, the latter of which it divides into three divisions; into the middle, which leads into the pyramid formed by the muscles of the eye, and which gives passage to the *nervus oculo-motorius abducens*, and the *ramus naso-ciliaris nervi trigemini*, and the *vena ophthalmica superiori*; in the upper division, which leads into the space between the upper part of the *bulbi*, with its muscles, and the wall of the orbit, and gives passage to the *nervus frontalis*, the *nervus trochlearis*, and the *nervus lachrymalis*; in the lower division, which enters the space between the globe and its muscles, and the wall of the orbit, and which gives passage to the *vena ophthalmica inferior*.

The tendinous ring gives origin to six muscles: the four *recti*, proceeding forward to be inserted into the anterior part of the eyeball, form a pyramidal space; the two other muscles, originating from the common tendinous ring, are the superior oblique, and the elevator of the upper lid. The space between the eyeball and the walls of the orbit is filled up with a loose connective tissue, quite rich in adipose matter. This connective tissue is condensed on some points, and constitutes sheaths for the muscles within the orbit, for the nerves and bloodvessels; it also forms fascia-like processes, which connect some of the parts within the orbit with the periorbita.

The *tunica vaginalis bulbi* (Bonnet's capsule, capsule of Tenon) is continuous with the optic nerve sheath, and envelops the eyeball loosely until it arrives in front of the equator, where it is perforated by the straight muscles of the eye; forward

of that point it is firmly connected with the sclerotica and
the conjunctiva, and proceeds forward as far as the border
of the cornea, where it ceases. It also sends out some processes to the edges of the cartilages of the upper and lower
lids, forms a connection between the conjunctiva and the *facia
tarsi orbitalis*; it also sends membranous bands to connect with
the *ligamentum canthi internum* and *externum*; and the *caruncula lachrymalis* (along with the *plica semilunares*), rests on
such a band, which has its practical importance in connection
with the operation for strabismus (Liebrich). The posterior,
loosely connected partition is smooth on its inner surface, so
as to allow the free rotatory movements of the globe. At the
point of perforation, it is intimately connected with the muscles, *so as not easily to permit a separation of the parts*, and it
sends off processes backward toward the optic foramen for
some distance as firmly connected sheaths to the muscles.
Anterior to the perforation, the tendons are not enveloped by
the capsule, but for some distance proceed without any envelope, until, just before their insertion, they expand between
the capsule and the sclerotica, to be inserted into the latter,
but also in close connection with the former. Anterior to the
line of insertion of the recti muscles, the capsule becomes much
thinner, and it is extremely difficult to detach it from the conjunctiva and the sclerotica, so intimately are the three membranes connected. Liebrich compares this anterior belt as a
lid of a half sphere to a half globe shell. The lid has a circular perforation on top, just the size of the cornea, at the border
of which it ends. The portion in front of the perforation of
the muscles is often named the *capsule of Tenon*, whilst the
loose posterior partition has also been known as *Bonnet's capsule*. In the anterior part of the capsule, it is intimately connected with the conjunctiva, and quite firmly up to an irregular circular line, which is known by the fact, that by eccentric
movements of the eyeball, a fold is produced at the line referred to. The formation of this fold prevents the corruga-
tion and projection forward of the conjunctiva, which would

otherwise take place, at the caruncula, on turning the eye inward. A complete continuation from one of those divisions
of the capsule into the other does not take place, as, at the line
formed by the fold referred to above, the border of the posterior division in part turns out to the border of the orbit on the
membranous expansions in the manner explained above. Back
of the equator, near the optic nerve, it is perforated by the
obliquus superior and a little further forward, on its outer and
posterior surface, by the *obliquus inferior*.

The Muscles of the Eye.

There are seven muscles within the orbit, the *levator palpebræ, rectus superior, rectus inferior, rectus internus, rectus externus, obliquus superior*, and *obliquus inferior*.

The four recti muscles, according to Alt's observations on
frozen eyes, proceed in a straight line, up to the greatest
diameter of the eyeball, and at the equator they first come in
close connection with the *bulbus oculi*, but still are outside of
the ocular sheath, and it is only near the insertion of their tendons into the sclerotica that the capsule is perforated. From
this point back to the equator of the globe, there is only a
rather loose cellular connection between the capsule and muscles. The muscles for some distance back have a sheath formed
from processes of the tunica vaginalis. After perforating the
ocular sheath, the tendons are naked for a short distance, but
they soon expand, between Tenon's capsule and the sclerotica,
to be inserted. The expansions of the tendons do not meet each
other, so as to form a continuous membrane, but are only connected in a manner by the tunica vaginalis bulbi. The tendons
of the recti muscles are quite short, are from $3\frac{1}{2}'''$ to $4'''$ broad,
with the exception of the *rectus externus*, which is $\frac{1}{2}'''$ nar-
rower. The common muscle plane for the *rectus internus* and
the *rectus externus* rests in the horizontal meridian; the common muscle plane of the *rectus superior* and the *rectus inferior*
is on the vertical meridian. This is, perhaps, not strictly cor-

rect, as the *inferior* inclines $\frac{1}{2}'''$ inward toward the nose. A line drawn through the points of insertion of the recti muscles, will strike a point between the equator of the eyeball and the border of the cornea, in the region of the *processes ciliares*, and its·vertical diameter is 8''', and its horizontal diameter 9''', whilst the vertical diameter at the equatorial plane is 10''' to $10\frac{1}{2}'''$, and the horizontal $10\frac{1}{2}'''$ to 11''' (Pilz).

Fig. 55.

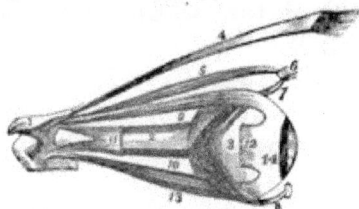

A view of the muscles of the eyeball, taken from the outer side of the right orbit. 1. A small fragment of the sphenoid bone around the entrance of the optic nerve into the orbit. 2. The optic nerve. 3. The globe of the eye. 4. The levator palpebræ muscle. 5. The superior oblique muscle. 6. Its cartilaginous pulley. 7. Its reflected tendon. 8. The inferior oblique muscle ; a piece of its bony origin is broken off. 9. The superior rectus muscle. 10. The internal rectus, almost concealed by the optic nerve. 11. Part of the external rectus, showing its two heads. 12. The extremity of the external rectus at its insertion, the intermediate portion of the muscle having been removed. 13. The inferior rectus muscle. 14. The sclerotic coat. (*From Smith's Anatomical Atlas.*)

The *superior rectus* (*musculus rectus superior, attollens oculi*), takes its origin in connection with the *musculus palpebræ*, from the upper part of the tendinous ring, and proceeds forward in the same direction as the *nervus opticus ;* yet in consequence of its direction to reach the upper end of the vertical diameter of the globe, it forms an angle with the optic axis of 20°. It is the most thin among the recti muscles, about an inch and a half in length, and is inserted $3\frac{3}{5}'''$ from the corneal border, with the inner side of its tendinous expansion 1''' nearer the cornea than its outer.

The inferior straight muscle (*musculus rectus inferior, seu deprimans oculi*), takes its origin from the lower part of the common tendinous ring, is thicker than the superior straight muscle, and 1''' to 2''' longer, and proceeds forward, bearing about the same relation to the optic nerve axis, as the rectus superior, and is inserted 3''' behind the lower border of the cornea, ½''' inward from the end of the vertical meridian, and with the inner border of its tendinous expansion 1''' nearer the cornea than the outer.

The *inner straight muscle* of the eyeball (*musculus rectus internus, seu adducens oculi*), arises from the common tendinous ring, proceeds forward parallel with the inner wall of the orbit, and is inserted 2½''' back of the inner border of the cornea, in the horizontal meridian. It is the thickest of the straight muscles, is about an inch and a half in length, and its tendinous expansion is quite broad, sometimes being divided into two tendons.

The outer straight muscle (*musculus rectus externus, seu abducens oculi*), arises by two heads from the common tendinous ring, and passes along the outer wall of the orbit, and in consequence of its direction outward is the longest of the recti, being 3''' longer than the rectus superior, and is next in thickness to the rectus internus. It is inserted further back from the border of the cornea than any of the other straight muscles, being 3½''', and at this point the lachrymal gland rests on it. Its muscle-plane is in the horizontal meridian.

There are two more muscles that run along the roof of the orbit, the *levator palpebræ superior*, and the *obliquus superior*.

The *musculus levator palpebræ superior* originates from the upper part of the common tendinous ring, in common with the *musculus rectus superior*, as a triangular flat muscle, is separated from the *rectus* in the region of the *bulbus*, and becomes broader, and passes out of the orbit below the *marga supra-orbitalis*, and behind the *ligamentum tarsi superioris*, and terminates in a flattened tendon, one half inch in breadth, to be inserted into the superior border of the tarsal cartilage, and into the fascia extending from that point.

The *musculus obliquus superior, seu trochlearis, seu patheticus*, originates also from the common tendinous ring, between the tendons of the *rectus superior* and the *rectus internus*, proceeds along the upper and inner angle of the orbit forward to the pulley or trochlearis beneath the internal angular process of the frontal bone, which is located about 5½''' above the horizontal middle of the cornea, 6⅜''' inward from the vertical diameter of the cornea, and about on the same plane with the base of the cornea (Pilz), over which its slender tendon passes, where it changes its direction, becomes broader, passing under the *rectus superior*. It is the longest and thinnest of all the ocular muscles, and by a tendon 3''' in length, with its convexity turned backward and outward, is inserted into the temporal side of the eyeball, with its inner end 3⅜''' to 4''' from the *nervus opticus*, whilst the outer end is inserted 6''' to 7''' further forward. The *musculus obliquus inferior* arises from a depression in the orbital edge of the superior maxillary bone, a little to the outer side of the lachrymal sac, runs along the floor of the orbit, beneath the *rectus inferior*, first for a distance of 3''' running backward and outward, and then turns upward and backward, forms a fibro-cellular connection with the *rectus inferior*, and passes between the globe and the *rectus externus*, and is inserted by a broad thin tendon, with short fibres, on the temporal side of the posterior part of the eyeball, with a line of insertion of 5''' in length, with its convexity directed forward and upward, the anterior end of which is 7''', and the posterior 2''' to 3''' from the optic nerve. The seven muscles of the orbit that have been described, one of which raises the upper lid, and the six others rotate the ball around the turning focus, consist of transversely striated muscular fibres. They are furnished with nerves from the *nervus oculo-motorius*, or third pair, with the exception of the *rectus externus*, which is supplied by the sixth pair, and the *obliquus superior* by the fourth pair of cerebral nerves. Consequently all are voluntary muscles.

Action of the Muscles of the Eye.

Although it is not the plan of the present treatise to enter extensively into the physiology of the parts described, a few words on the actions of the muscles above considered seem to be necessary here ; and first a few definitions must be given, in doing which we shall follow Pilz. The optic axis (*a b*, Fig. 56) is the extension of the lens axis forward to the cornea, as

Fig. 56.

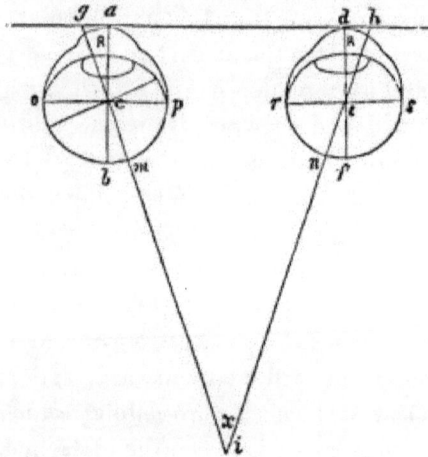

well as to the back part of the eye, and is $10\frac{2}{3}'''$ to $11'''$ in length. The points that are touched by the horizontal line along the surface of the eyes, are the *poles*, of which there are two, the anterior (*a d*) and posterior (*b f*) poles of the eyeball. The *equatorial plane* is a line vertical to the optic axis (*o p, r s*), through the greatest diameter of the eyeball, and is located $\frac{2}{3}'''$ to $1'''$ nearer the posterior pole than to the anterior, and its longest diameter is neither horizontal nor vertical, but extends from the nasal side above to the temporal side below.   It measures the same as the optic axis, whilst the vertical diameter is $\frac{1}{5}'''$ to $\frac{1}{4}'''$ shorter. The circumference around the equatorial plane is the *equator*, and the sections of the globe thus made, are the anterior and posterior *hemispheres*. The *meridional planes* are drawn through the optic axis, and cut the globe into lateral segments ; and those curved lines on the surface of the globe, centering in the meridional plane, are the *meridians*.

The *optic nerve axis* is a line drawn through the middle of the optic nerve entrance to the middle of the optic axis, and from thence is extended to the anterior surface of the cornea, 1½''' outward from the anterior pole (g, h), and the posterior end of the same axis is just the same distance to the inner side of the optic axis. It cuts the optic axis at an angle of 20° 32'.

The eye must be considered as a globe, with a fixed centre (*centre of motion—Drehpunct*) around which latter it rotates without change of locality. A *muscle plane* is the direction in which a muscle exerts its force, and is obtained by drawing a line from the middle of its origin to the middle of its insertion, and a line which stands perpendicularly upon this plane in the turning-point, is the *axis of turning* (*Drehungsachse*). Consequently the axis of turning of the *musculus rectus externus* and *internus*, is vertical, corresponding with the vertical diameter of the eye, leaving the vertical meridian unchanged. The common muscle plane of the *rectus superior* and the *rectus inferior* does not coincide with the vertical meridian, but for the simplification of the matter, they are so considered by authors. It inclines itself a little from behind and forward, and within outward. This plane, which cuts the antero-posterior axis of the eye, under an angle of 20 degrees, does not represent a meridian of the globe, and does not contain the centre of the eye, but leaves it outward. The perpendicular axis of turning to this plane, forms, with the horizontal diameter, an angle of 20 to 25 degrees, and with the antero-posterior an angle of 70 degrees (Wecker). These two muscles contracting separately will draw the centre of the cornea upward and downward, but from the anatomical conditions described, the superior and inferior straight muscles, besides each one drawing the centre of the cornea to its own side, when acting in concert, they will cause a slight deviation of the cornea inward. The *rectus superior* contracting, will draw the superior extremity of the vertical meridian inward, the *superior* extremity of this meridian being the one always noted in determining the ocular movements, whilst the *rectus inferior*, in contracting, draws the inferior extremity of the same meridian inward, and consequently the superior extremity is inclined outward, and we

8

say the vertical meridian is inclined outward. The contraction of this group of muscles is not always the same as regards the position of the centre of the cornea, as well as the inclination of the meridian, and varies according to the position of the antero-posterior axis of the eye (optic axis) at the moment of contraction. If the centre of the cornea is directed outward in such a manner, as that the axis of turning of the group forms with the optic axis an angle approaching closely to 90 degrees, and that the muscle plane tends progressively to confound itself with the vertical meridian, the muscular contraction will necessarily produce its effects upon the deviation above or below the centre of the cornea, whilst the inclination of the meridian will be very little affected. Inversely, if the optic axis is directed inward, and forms with the axis of turning a less and less open angle, and if the muscle plane forms with the vertical meridian an angle more and more open, the influence of the contraction of the muscles of this group will have a decided influence upon the inclination of the vertical meridian, whilst the deviation in height will be quite feeble (Wecker). For the sake of convenience in studying the actions of the oblique muscles, a voluntary mistake is necessitated in adopting a common muscle plane. This plane does not accord with the equatorial plane of the eye, but turns aside by its internal circumference forward and by its external circumference backward. The centre of the eye is located in this muscle plane, and which deviates more from the equatorial plane, forward and inward, then backward and outward. Its axis of turning does not, then, accord with the optic axis, but according to Graefe forms with it an angle of 35 to 40 degrees. The anterior extremity of this axis ends outward from the anterior pole of the eye. The action of each of the oblique muscles is trifold as in the group preceding this one. The superior oblique draws the centre of the cornea downward and outward; but its effect is to draw the superior end of the vertical meridian inward, and consequently to incline the meridian inward. The inferior oblique draws the centre of the cornea upward and outward, and draws the inferior extremity of the vertical meridian inward, which inclines this meridian outward. The effect that will be produced by the action of these muscles, whether on the deviation of the height of the corneal centre, or whether it be on the inclination of the meridian, will vary much according to the angle that the optic axis will form with the axis of turning; or in other words, according as the anterior pole of the eye will be directed inward or outward. The effect that the oblique muscles produce upon the height of the cornea, is the more decided, when the eye is strongly carried outward; for then the axis of turning of the obliqui forms with the optic axis an angle more and more approaching a right angle. On the contrary this angle diminishes by the approach of the two axes, when the cornea is directed outward, and for the reason that the muscular contraction then produces its effect nearly exclusively upon the meridian which it deviates, and not upon the height of the centre of the cornea. In looking over what has been said on the actions of the above muscles, we find that the obliqui are the antagonists of the recti. The consentaneous action of the recti, draws the eye backward into the orbit. The obliqui pull the globe out. The former are retractors, whilst the latter are protractors. The division of the recti, would enable the obliqui to draw the eye forward and make it protrude from the socket; to divide the obliqui, would cause the recti to retract the ball within the orbit. The action of the recti is greatest on the height of the cornea, when the optic axis is directed outward. The reverse is true of the obliqui, which have their maximum effect upon the vertical meridian when the cornea is turned out, whilst the contrary is true of the recti superior and inferior. The combined action of the *rectus superior* and the *obliquus inferior* deviate the cornea upward, and these muscles are antagonists as relates to the deviation laterally and to the inclination of the meridian. The same relations exist between the *rectus inferior* and the *obliquus superior*.

Now briefly as regards the actions of the individual muscles:

In looking straight forward, all the muscles are in a state of equilibrium, and the vertical meridian is perpendicular. In looking inward on the horizontal plane, the only muscle concerned is the *rectus internus,* which leaves the vertical meridian unchanged. In looking outward on the horizontal plane the *rectus externus* is the only muscle in action. Hence in moving both eyes on the horizontal plane, the only muscles in action are the *rectus internus* and *externus.* In looking directly upward the vertical meridian remains vertical. To bring about this movement the muscle chiefly concerned is the *rectus superior.* Now this muscle acting singly would also incline the vertical meridian inward, and consequently it must associate itself with an antagonist in order to keep the meridian vertical, which we find to be the *obliquus inferior.* In looking directly upward with both eyes, four muscles then are in a state of activity. So in looking directly downward, the vertical meridian remaining vertical, the *rectus inferior* associates with itself the *obliquus superior.* In looking upward and outward, the vertical meridian is inclined outward, and the muscles chiefly concerned in bringing about this result are the *rectus externus* and the *rectus superior.* But in order to counteract the tendency which the *rectus superior* has to incline the vertical meridian inward, the *obliquus inferior* is brought into action, to draw on the inferior extremity of the meridian, and assist in inclining outward. Then in looking upward and outward with both eyes, six muscles are in action. The *rectus externus* simply turns the eye outward, the *rectus superior* rotates it upward, and the *obliquus inferior* inclines the meridian. In looking downward and outward, it has been determined that the vertical meridian is inclined inward. The *rectus externus* rotates the eye outward, and the *rectus inferior* rotates it downward, with the vertical meridian slightly inclined outward. To correct this the *obliquus superior* is brought into action to correct the inclination of the meridian. Then, we find, in looking downward and outward, six muscles act. The *rectus externus* to rotate outward, the *rectus in-*

ferior to rotate downward, and the *obliquus superior* to incline the meridian inward. In looking upward and inward, the vertical meridian is inclined inward. The *rectus internus* simply rotates the cornea inward; the *rectus superior* rotates the eye upward, and also inclines the vertical meridian inward, and with the greater force, the optic axis being directed inward. But a third muscle is necessary to modify the power of the two muscles named, in their tendency to incline the meridian inward; this we find to be the *obliquus inferior*, which partly counteracts the action of the *rectus superior*, at the same time that it assists the latter in elevating the centre of the cornea. To look upward and inward, then, employs six muscles, the *rectus internus*, to rotate the eye inward; the *rectus superior*, to rotate the eye *upward*, and incline the meridian; and the *obliquus inferior*, to assist in rotating the centre of the cornea upward, and also to modify the inclination of the meridian. In looking downward and inward, the vertical meridian is inclined outward. The rotation inward is effected by the *rectus internus*, and the centre of the cornea is rotated downward by the *rectus inferior*, which at the same time finds the eye in a favorable position for inclining the meridian outward. We shall find, again, that a third muscle is required to modify the excessive influence of the *rectus inferior*, which we find to be the *obliquus superior*. The action of both eyes for vision downward and upward, requires the activity of six muscles: the *rectus internus*, to rotate inward; the *rectus inferior*, depressor and deviator of the meridian; and the *obliquus superior*, a depressor of the corneal centre, and modifies the deviation of the meridian. It will be perceived that when the eye is moved either strictly *vertically* or *horizontally*, the meridian remains vertical; in the positions outward, it is the obliqui that determine the inclination of the meridian; in the positions inward, it is the *rectus superior* and *inferior* that incline the meridian. In these different positions, the muscles act with the greater facility, the more the optic axis confounds itself with their axis of turning.

The Optic Nerve (Nervus Opticus.)

The optic nerve is somewhat tortuous in its passage from the optic foramen to the globe, forming a curvature outward and downward. Its length from the eyeball to the optic foramen is 13''' to 14''', which is more than the distance of the posterior pole from the same foramen, which is only 12'''. This is owing to the fact that the optic nerve is inserted to the inner side of the posterior pole. It is surrounded by the posterior ciliary arteries and the ciliary nerves, which run along close to its sheath and accompany it through the loose fatty tissue to the *bulbus oculi*. During its course it receives, and, in its axis, carries within it, the *arteria centralis retinæ*. It is 2''' thick, and has a considerable constriction as it passes through the sclerotica. (See Fig. 1.)

The optic nerve is a nerve of special sense, and its exclusive office is to conduct to the nervous centre the luminous impressions made upon the retina. It runs a long course in the encephalon before its termination in the eye. It is in immediate connection with the optic tracts of the brain. What is said below in reference to the cerebral origin of the optic nerve is mostly translated from Galezowski.

The optic tracts are white, and rather broad, and proceed directly from the *corpora geniculata*, external and internal; they are in contact with the *cerebral peduncles*, around which they pass horizontally, and proceed forward to form the *chiasma*. In half their course they are in contact with the cerebral peduncles. They are in relation with the cerebral peduncles above and within; below and outward they are free. Further forward they are in contact with the *membrana perforatus* and the sphenoid bone.

The optic tracts proceed, as stated above, from the *corpora geniculata*, located on the inferior and posterior surface of the optic thalami. The internal is smaller than the external, but is more prominent. On its free surface it is enveloped by a thin white layer, which is prolonged backward to the posterior or quadrigeminal tubercle (*testes*). A nucleus of gray substance is found in it.

The external *corpora geniculata* form an oblong eminence, which passes around the posterior and inferior extremity of the optic thalami, and by means of a small white medullary band, communicate with the anterior *quadrigeminal tubercle* (*nates*). Its free surface is also white, but of a more murky color. The existence of gray nuclei in these bodies leads to the opinion that they are not simply conductors, but that they perform some special act connected with vision.

The *tubercula quadrigemina* are formed by four separate eminences divided by grooves. The two anterior projections are called *nates*, and the posterior *testes*. They are in front of the cerebellum and above the cerebral peduncles, with which they form adhesions.

In structure, they are very complex, and are covered externally by a thin, white substance. In the posterior tubercle are found round nuclei, of a reddish-gray color, more dense than the cortical part.

Another portion of gray matter, separated by the two nuclei described, serves to connect the tubercles of the opposite sides. Beneath these pass the white medullary fibres pertaining to the band of Reil, and to the cerebellar peduncles.

It then seems that the *tubercula quadrigemina* are in direct, or indirect, connection with the several parts of the brain, as with the cerebellum, the medulla oblongata, and the spinal cord. Thus, there is given off from each posterior tubercule a small white band of medullary matter, which passes in front of the triangular lateral fascicule of the isthmus, to end in front in the *corpora geniculata interna*. There is another prolongation forward and outward, which proceeds directly to the *corpora geniculatum externa*. The pineal gland is attached to the *tubercula quadrigemina* by four white fasciculi, as also by the choroid plexus. The valve of Vieussens communicates with the posterior extremity of the *tubercula*, by its filament; the cerebellum communicates with the *tubercula quadrigemina* through the *processus cerebelli ad testes,* which take their origin in the white nuclear substance of the cerebellum, cross above the inferior peduncles of the same organ, and pass beneath the band of Reil and in the *tubercula quadrigemina.* Between the medulla oblongata and the *tubercula quadrigemina,* there is a communication through the antero-lateral fasciculus of the bulb or band of Reil.

M. Schröder van der Kolk has demonstrated that the anterior roots terminate wholly in the anterior cells, forming a group; from these proceed the ascending or encephalic fibres, which form the anterior and lateral cords. The posterior roots contain two orders of fibres, the cerebral and the reflex; the first ascend directly into the brain; the others end first in the cells of the posterior cornu, and thence proceed to the brain. Hence, it is inferred that the superior crura of the cerebellum form a continuation with the posterior medullary fascicule, and that the fibres of the band of Reil are, according to Longet, the continuation of the antero-lateral fascicula of the spinal cord.

According to this, then, the *tubercula quadrigemina* are in relation both with the posterior or sensitive, and with the antero-lateral or motor columns of the spinal cord, which explains why affections of the latter so frequently cause amaurosis (Galezowski).

The *chiasma* or *commissure* is nearly quadrilateral, and is situated on the olivary processes of the sphenoid bone. It is bounded in front by the *lamina cinerea,* by the *tuber cinereum* behind, and on either side by the anterior perforated space.

In the chiasma, the optic nerves of the two sides partially cross. It contains two kinds of fibres (see Fig. 57): the external, which proceed to the external part of the retina, and which do not decussate; the central, which cross in the chiasma; and the inner fibres of the optic tract of one side cross over to supply the inner side of the retina of the other side, and *vice versa.*

Another class of fibres,—the internal—do not decussate,

but those on the anterior surface of the chiasma pass from the retina of one side to the retina of the other side, without crossing, and are called the *inter-retinal* fibres; those on the posterior surface of the chiasma do not cross, but pass from one optic tract to the other, and are called the *inter-cerebral* fibres. (See Fig. 57.)

FIG. 57.

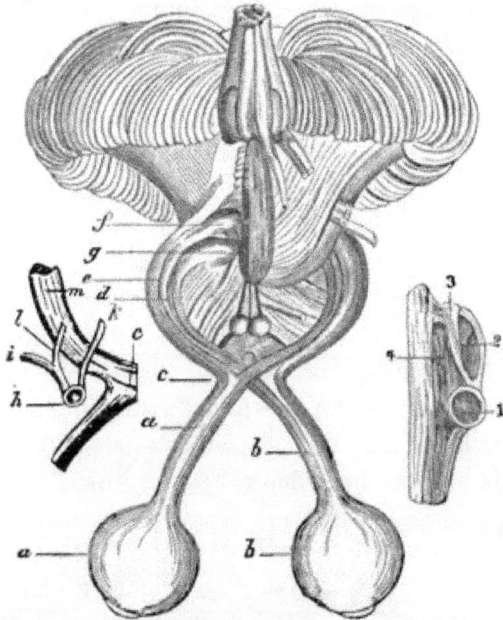

a, a. External fibres of the optic nerve and of the retinæ coming from the corresponding hemispheres. b, b. Internal fibres of the same nerve which originate in the opposite hemispheres. c, c. Chiasma, with inter-crossing of the optic fibres. d, e. Optic tracts. f. Testes. g. Nates. h. Internal carotid. i. Middle cerebral artery. k. Posterior communicating artery. l. Anterior optic artery.

1. Central nucleus of the testes. 2. Gray substance of the nates. 3. White band which separates the nates in two parts. Below the gray part 4, are seen parallel white striæ, which are the continuation of the processus cerebelli ad testes. (*From Galezowski.*)

The former constitute a communication between the retinæ of both eyes; the latter becomes continuous with the optic tracts. Also see Fig. 63, of which o, o, is the *commissura arcu-*

ata anterior; r, r, the *commissura arcuata posterior; p, p, q, q,* the *commissura cruciata.* Up to the chiasma the fibres are medullary; at the commissure the optic tracts present modifications that are important in a physiological point of view. Each one, before crossing over, receives on its superior·face a large fascicule of gray fibres, emanating from the gray mass which covers the internal surface of the optic thalami. (Wecker.) At the chiasma the pia mater forms a sheath for the optic nerves, which accompanies them to their insertion into the globe. It sends numerous processes inward in a manner presently to be described. At the optic foramen each optic nerve receives a strong outer sheath, generally considered as a continuation of the dura mater. It is believed by some (Stellwag) to be derived from the periorbita; but the weight of testimony seems to preponderate in favor of its origin from the dura mater. According to Donders (*Archive für Ophthalmologia,* Band i, Ab. ii, Bl. 82), the optic nerve, as regards the structure of its sheath and its processes between the individual nerve fasciculi, differs from other nerves.

In ordinary nerves Donders found a firm fibrous texture as a common sheath (see Fig. 58), which sends extensions inward (*a'*), and is crossed by a loose cellular tissue (*b*), which (the

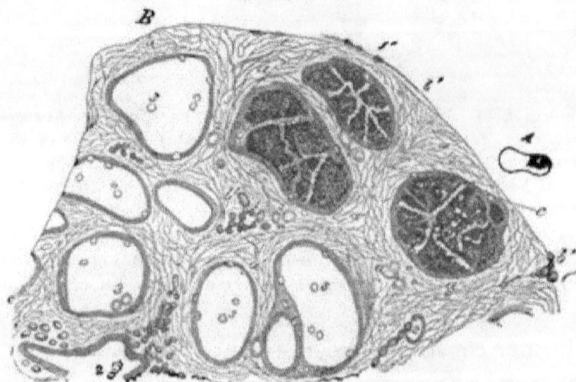

Fig. 58.

Transverse section of the ischiatic nerve of a man. Magnified 50 diameters.
(*From Donders.*)

latter) separates the tertiary fasciculi, and the larger branches of vessels (2, 2), as well as inclosing fat-cells (*i*); besides, there is around each secondary nerve-bundle a thin, firm, fibrous, lamellated enveloping membrane (*c*), the *neurilemma proprium*, which has no connection with the outer, common sheath;

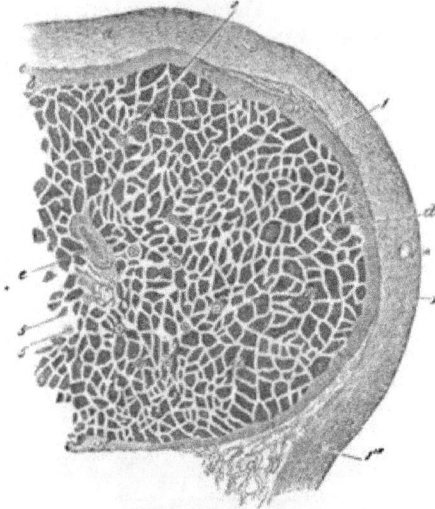

Fig. 59.

Transverse section of the optic nerve 1½ mm. from the sclerotica. (*From Donders.*)

whilst some loose cellular tissue divides the secondary bundles into the primitive, and takes up the smaller vessels (3).

In the optic nerve, on the contrary, we find in the longitudinal section (Fig. 60), as also in the transverse section (Fig. 59), two firm fibrous sheaths, an outer and thicker (*a*), and an inner and thinner (*b*), both rich in connecting elastic elements. Between these two sheaths there is a loose, connective tissue (*c*), in consequence of which the inner sheath, firmly connected with the nerve, can slide on the outer sheath. The outer sheath and the loose connective layer between the sheaths do not come in contact with the nerve-bundles at any point. The inner sheath sends in firm fibrous extensions, which separate the nerve-bundles (*c*), and there is no loose connective tissue

either between the secondary or tertiary fasciculi. The elastic
elements here are less developed, and seem only extensions with
connecting small elongated granules. The outer sheath (a',
Fig. 60), as seen in the longitudinal section, is lost in the outer
two-thirds of the sclerotica (a' b'). The inner sheath (b), on

Fig. 60.

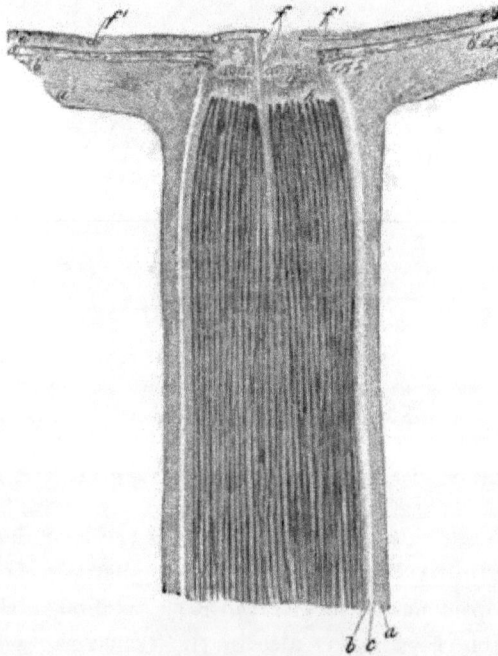

Longitudinal section of the optic nerve and the tunics of the eye. Magnified
10 diameters. (*From Donders.*)

the contrary, encircles the nerve-stem close to the *choroidea*,
with which some of its fibres undoubtedly connect, whilst the
others turn out immediately beneath the choroid, and are
blended with the inner sclerotica. From this inner portion of
the sclerotica a number of elastic elements are given off, which
proceed between the individual nerve-bundles of the optic nerve
to form the so-called *lamina cribrosa* (g). Donders asserts, dif-

fering from Kölliker and Heinrich Müller, that the fibres are
in contact with only a very small part of the choroidea, and
that within the eye, in fact until within the retina itself, the
nerve-bundles are separated by the inter-fascicular tissue.

There are in the optic nerve a great number of small arterial
and venous branches. Those on the outer sheath are partly
continuous with those of the sclerotica; those in the loose cellular tissue, between the outer and inner layers of the sheath
(which, immediately behind the *lamina cribrosa*, becomes continuous with the inner sheath, and at this point, where it is
already within the sclerotica, it is thickest), pass on the inner
sheath, and on the processes separating the nerve-bundles, and
are distributed on the *lamina cribrosa*, and some pass into the
papillæ nervi optici, and surrounded by nervous substance, are
distributed in the retina itself. This is true especially of a
few small branches in the immediate neighborhood of the *vasa
centralia*.

As regards the distribution of the central vessels themselves,
every branch is completely surrounded by the optic fibres, not
one reaching the *membrana limitans*. Capillary vessels are
seen on every part of the retina, except in the bacillar or rod-
layer, and in the granule-layer, contiguous to it. As the
bloodvessels are placed on the inner part of the retina, the
inquiry presents itself whether in the *papilla nervi optici* the
vessels come to the surface. Donders satisfied himself that in
nearly all cases they are also surrounded by nervous fibres, and
that they seldom come in contact with the *membrana limitans*.
The vessels of the retina are an independent system of vessels,
having no communication with any other system of vessels.
At the anterior border of the retina they end in loops.

The Arteries.

The eyeball is (and its appendages mostly) supplied with
blood from the *ophthalmic artery*, which arises from the *internal
carotid artery*, just as the latter vessel is emerging from the

cavernous sinus, on the inner side of the anterior clinoid process, and enters the orbit through the optic foramen, below (*a*, Fig. 61, left hand) and on the outer side (*b*) of the optic nerve. It then crosses over the nerve to the inner side (*c*),

Fig. 61.

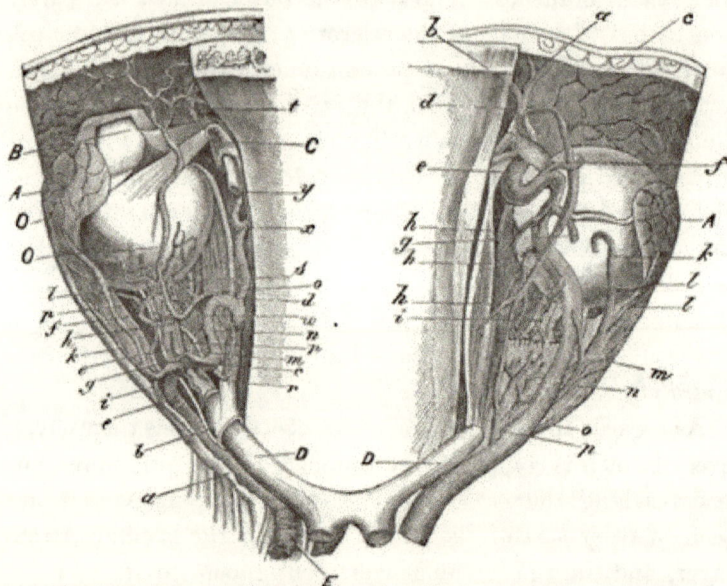

Arteries of the eye, left-hand figure; veins, right-hand figure. (*From Pilz*.)

leaves the muscle-cone, passes beneath the *musculus obliquus superior*, to the inner wall of the orbit, forward (*d*), to the inner angle of the eye, where it divides into two terminal branches, the *arteria frontalis*, and the *arteria dorsalis nasi*. The first branch given off from the ophthalmic is the *arteria lachrymalis* (*i*), which arises near the optic foramen, and accompanies the lachrymal nerve along the upper border of the external rectus muscle, and is distributed to the lachrymal gland. The terminal branches that escape from the gland are distributed to the upper lid and to the conjunctiva, and anastomose with the *arteria palpebralis*.

The lachrymal artery gives off one or two malar branches,

one of which passes to the temporal fossa, through a foramen in the malar bone, and anastomoses with the deep temporal arteries. The other passes to the cheek to anastomose with the *arteria transversa faciei*. The next branch given off from the ophthalmic is the largest, the *arteria supraorbitalis* (*s*), and is given off above the optic nerve. It passes forward with the frontal nerve, and passes above the muscles, between the *levator palpebræ* and the *periorbita ;* passing through the *foramen supraorbitale*, it divides into a superficial and a deep branch, which supply the muscles and the integument of the forehead, and the pericranium, and anastomose with the *arteria temporalis*, the angular branch of the *arteria faciei*, and the artery of the opposite side.

The *arteria ethmoidalis anterior* (*x*) *et posterior* (*w*) are given off at the point where the ophthalmic artery reaches the inner wall of the orbit (*d*). The former accompanies the nasal nerve through the anterior ethmoidal foramen, supplies the anterior ethmoidal cells and frontal sinuses, enters the cranium, and divides into a meningeal branch, to supply the dura mater, and a nasal branch, which passes through an aperture in the cribriform plate into the nose.

The *arteria palpebralis superior et inferior* arise opposite the *trochlea*, from the *arteria ophthalmica*, encircle the eyelids near their free margin, between the tarsal cartilage and the *musculus orbicularis;* they form free anastomoses, as with the infraorbital artery, and with the orbital branch of the temporal artery, and send off a twig to ramify on the nasal duct.

The *arteria frontalis*, one of the terminal branches of the *arteria ophthalmica*, passes from the orbit at its inner angle to be distributed to the muscles and integument of the forehead, and anastomoses with the *arteria supraorbitalis*.

The *arteria nasalis*, the other terminal branch of the *arteria ophthalmica*, passes from the orbit above the tendo-oculi, gives off a branch to the *saccus lachrymalis*, divides into two branches, which anastomose with the *arteria angularis*, and sends a branch to supply the surface of the nose, the *arteria dorsalis nasi*.

The *arteriæ ciliares*, are divided into *arteriæ ciliares breves*, *arteriæ ciliares posterior longæ, et arteriæ ciliares anticæ*.

The *arteriæ ciliares breves* are from twelve to fifteen in number, arise from the *arteria ophthalmica*, or its branches, and penetrate the sclerotica around the optic nerve, to supply the choroid, as heretofore explained.

The *arteriæ ciliares longæ*, two in number, perforate the sclerotica on the inner and outer sides of the optic nerve entrance, to supply the ciliary body and the iris, as explained in the description of the choroidea.

The *arteriæ ciliares anticæ* are a number of muscular branches that penetrate the sclerotica, close to the corneal border, and send branches to the ciliary body and the great arterial circle of the iris.

The *arteria centralis retinæ*, the smallest branch given off from the *arteria ophthalmica*, penetrates the optic nerve near the sclerotica, and passes along the axis of the nerve into the eye, to be distributed, as elsewhere described.

There are two *muscular* branches, a superior and an inferior, which supply the muscles of the eyeball. The superior supplies the *levator palpebræ, rectus superior*, and the *obliquus superior*. The inferior branch supplies the external and inferior *recti* and *obliquus inferior*. This branch furnishes most of the anterior ciliary arteries. These branches expand beneath Tenon's membrane, and will be more minutely described after the description of the veins of the orbit.

The Veins.

The *vena ophthalmica*, in its branches, is very similar to the *arteria ophthalmica* (Fig. 61, right hand figure). It begins at the inner angle of the eye, where it anastomoses with the anterior facial vein (*d*), and which collects the *vena frontalis* (*a*), the *vena dorsalis narium* (*b*), and the *vena supraciliares* (*c*). It proceeds backward along the inner wall of the orbit, enters the cavity formed by the ocular muscles, but does not pass out through the *foramen opticum*, but through the *fissura orbitalis superior* into the cranial cavity, and empties itself into the *sinus cavernosus.*

Outside of the muscular pyramid, the *vena ophthalmica* is joined by the *venæ palpebralis* (*f*) and the *vena sacci lacrymalis;* and inside of the muscular cone it is joined by the veins from the muscles (*h, h, i, m, n*), which are ordinarily joined, immediately behind the corneal border, by the anterior ciliary veins. It is also joined by 4 to 6 branches of the *vasa vorticosa* (*g, k*), and a few branches of the posterior ciliary veins, which Leber correctly asserts have their origin only in the sclerotica. It is said in the text-books that there are two long posterior ciliary veins corresponding to the long ciliary arteries. Leber asserts that they cannot be found in the eye; which coincides with the observations of the author. It is also joined by the *vena glandulæ lacrymalis* (*l, l*) and *vena centralis retinæ,* also by the *vena supraorbitaria* (*p*); which receive some branches from the ocular muscles; and sometimes this vein enters the cavernous sinus singly.

The sclerotica receives its supply of blood from the posterior and anterior ciliary arteries; and the veins carrying off the blood have been described above. The small vessels form on the sclerotica a vascular network of large meshes, which have the peculiarity, that almost every arterial branch has a vein accompanying it on each side. The veins anastomose much more frequently than the arteries. The larger branches of the arteries are tortuous in their course. From this network of vessels a fine network of capillaries originates. Those derived from the straight muscles, the anterior ciliary arteries, pass through the tendons of the muscles. The branches of the scleral network derived from the anterior ciliary arteries, anastomose with the posterior network derived from the posterior ciliary arteries.

As far as the sclerotica is covered by the conjunctiva we must distinguish its vessels from the deep layer of the anterior ciliary vessels. Both are at their periphery wholly isolated,

as the conjunctiva derives its vessels from the superior and
inferior palpebral and the lachrymal arteries. It is only near
the corneal border where the union between the two systems
of vessels begins, and at this point the ciliary vessels send out
numerous vascular loops to anastomose with those of the conjunctiva, which run backward in a radiary manner, and supply
the innermost zone of the sclerotical conjunctiva with twigs
for a distance of 2 to 3 mm. The looped network which overlaps the border of the cornea is also derived from branches of
the anterior ciliary arteries.

The anterior ciliary arteries pass through the tendons of the
muscles to the surface of the sclerotica, generally two branches
to a muscle (one to the *rectus externus*), and run in a very tortuous manner toward the corneal border, where they send
branches to the sclerotical capillary network ; but by far the
larger proportion of branches penetrate the sclerotica in a
manner elsewhere described. The branches given off to the
capillary network of the sclerotica are very fine, and most of
them are given off from the anterior ciliary arteries, as they
run on the surface of the sclerotica ; a small number, however, proceed from the branches penetrating into the sclerotica.

Nearer to the corneal border their terminal branches, and
sometimes also some of the larger branches, bend laterally,
and form bow-shaped connections with each other, and from
which new branches are given off and proceed forward and
outward. From these bow-shaped communications proceed
the vascular loops which pass into the conjunctiva, in the
annulus conjunctivæ, which are the direct communication between the vessels of the sclerotica and the conjunctiva ; also
the branches are given off which form the looped vascular
network at the border of the cornea. The former exist in
the *annulus conjunctivæ*, side by side, at a distance of $\frac{1}{4}$ to $\frac{1}{2}$
millimetre; at first they run a short distance forward, bend
outward in a bow-shaped manner, and reach the conjunctiva,
in which they run for some distance directly backward, send
off some twigs, and unite with the arteries coming from the

periphery of the conjunctiva. The looped network of the
cornea is supplied by the terminal branches of the ciliary
arteries. They run forward on the sclerotica in a straight
manner, and by continual dichotomous divisions, and free
anastomoses, form a rather large-meshed and sharp-angled
vascular network, characterized by its great fineness. It
overlaps the border of the cornea to a distance varying with
the breadth of the *limbus conjunctivæ*, being broader above and
below than on the inside and outside. Its terminal branches
bend over in a curved manner, and pass over into the commencement of the veins, when they become wider, the venous
arm of the angle being at least double that of the arterial.
The looped vascular network of the corneal border also presents a very complicated capillary network (the veins of which
pass into the anterior ciliary veins), which is only indirectly
connected with the vessels of the conjunctiva. The anterior
ciliary veins are ordinarily more numerous than the arteries,
are less tortuous in their course, anastomose more freely, but
in their general course they are analogous to the arteries.
They collect the veins from the vascular network of the corneal border, from the capillary network of the sclerotica, from
within the globe, from the ciliary plexus and ciliary muscles,
and from the vessels forming the connecting loops of the conjunctiva. The veins proceeding from the looped vascular
network of the corneal border form a network of polygonal
meshes, which extends backward on the sclerotica, and forms
a zone of 2''' to 3''' in breadth, around the corneal border.
This is generally called the *episcleral* venous network, as it
rests on the sclerotica. This is, during life, much injected
during intra-ocular inflammation. The veins of the conjunctiva, collecting the blood by the anterior ciliary veins, have
the same direction as those described above as coming from
the connecting vascular loops. Like those, they are located
in the *annulus conjunctivæ*, and pass in a looped or sling-like
manner to the conjunctiva, on which they run back to form
connections with the peripheral conjunctival veins.

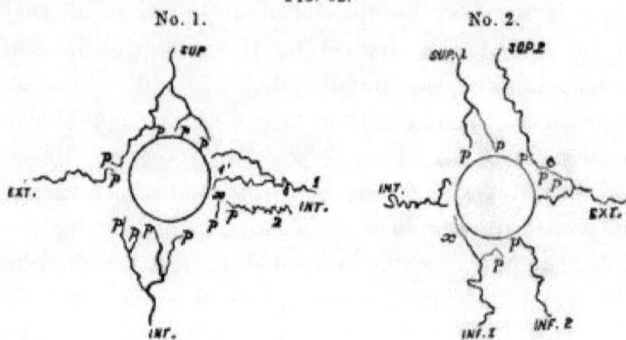

Fig. 62.

No. 1.

No. 2.

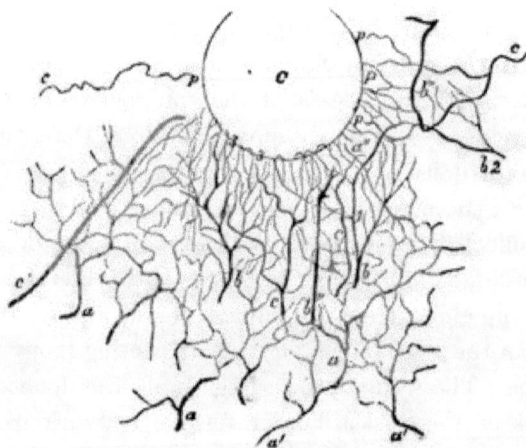

No. 3.

No. 4.

No. 5.

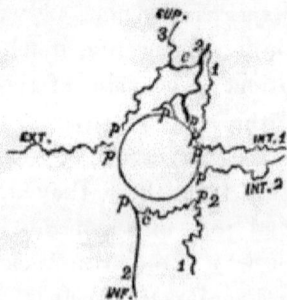

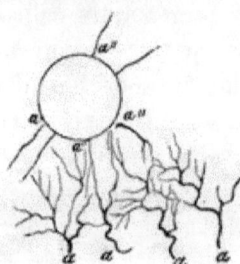

No. 6.　　　　　　　　　No. 7.

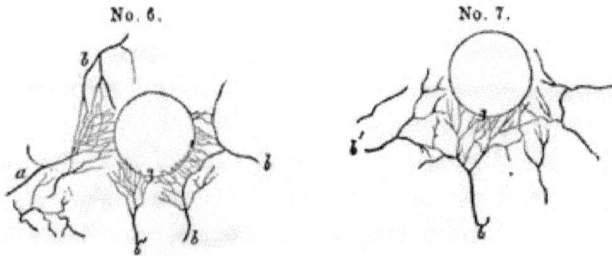

Explanation of the Figures.

No. 1. Right Eye. Anterior ciliary arteries of a man aged 37 years.

Sup. Superior, very deeply situated, with four perforating points, *p, p, p, p.*

Inf. *Inferior superficial,* also having four perforating points.

Int. *Internal,* 1, with the inferior ramus 1' hidden in *x,* under a pinguecula ; 2, very fine branches go out separately from the right inner muscle.

No. 2. Left Eye. Anterior ciliary arteries of a man aged 37 years. (Signification of figures and letters same as above.)

c. Anastomoses between the external and superior external branches, with perforating points, *p, p, p.*

x. One of the internal branches, which seems to terminate in an episcleral vessel.

Int. Internal, with its main tract very sinuous.

No. 3. Sketch of the vessels on the inferior portion of the external part of the eye.

c. Cornea.

a, a, a. Conjunctival vessels, being clear red, movable.

a', a'. Conjunctival vessels proceeding from the palpebral conjunctiva.

a", a". Anterior conjunctival vessels, of which a branch forms an anastomosis with an episcleral vessel at *d'.*

b. An episcleral vessel, connected by a loop with *b* 2.

b'. An episcleral vessel, with arborescent ramifications.

b". A deep-seated vessel, going out of the sclerotica.

b'". Episcleral network.

3, 3. Fine network surrounding and covering the border of the cornea.

c, c. Anterior ciliary arteries perforating the sclerotica at *p, p, p.*

c'. A perforating artery situated in the conjunctiva.

No. 4. Anterior ciliary arteries of a boy 13 years old.

Sup. Superior. 1, internal, very small, proceeding perhaps from 2, the middle dividing in three branches, and anastomosing at *c* with 3, the external.

Ext. External, a very fine trunk.

Inf. Inferior, 1 the internal, after a sinuous course returns to 2, the external ; this last, toward the point of junction, *c,* is situated in the conjunctiva, is rectilinear, and of clear red color.

Int. Internal. 1, the superior ; 2, the inferior ; *p, p, p,* points where the arteries disappear in the sclerotica.

No. 5. Conjunctival vessels.

a, a. Posterior.

a", a". Anterior, which near the cornea is bent partly into an episcleral vessel.

No. 6. Episcleral vessels.

b, b. Forming a loop.

b'. Arborescent.

a. A vessel situated in the conjunctiva, and communicating with the episcleral vessels *b.*

No. 7. Episcleral vessels dilated by friction.

b', b'. Forming a loop.

3. Fine network around the cornea.

(*From Donders.*)

The arteries that supply the peripheral part of the conjunctiva are the superior and inferior palpebral and the lachrymal. In a similar arrangement the veins form an arborescent network of irregular meshes.

The above observations on the vessels of the sclerotica and the sclerotical conjunctiva, from Leber, and based on observations made on dead eyes, finely injected, are mainly substantiated by the observations of Van Woerden and by Donders, on living eyes, by means of the microscope of Amici, adapted to that of Liebreich. Van Woerden calls those vessels of the conjunctiva which surround the border of the cornea, the *anterior conjunctival vessels*, in contradistinction to the *posterior conjunctival vessels*, which have their origin from the palpebral vessels. In the peripheral part of the conjunctiva there is but little communication between the conjunctival and ciliary vessels. This communication becomes more free toward the border of the cornea, and in the *limbus conjunctiva* the fine vessels pass over into the conjunctiva freely in the manner above described. It follows that in the neighborhood of the corneal border the conjunctiva is supplied from the anterior ciliary arteries, and that under certain conditions the conjunctival vessels may receive blood from, or carry it to, the intraocular circulation. Donders says that the vessels of the interior of the eye, arteries, and veins, communicate with those of the exterior, not excluding those of the conjunctiva.

Nerves of the Eye.

The second pair of cranial nerves, the *nervus opticus*, being a nerve of special sense, has been described above.

The *ganglion ciliare* (Fig. 63, *k*) is a small quadrangular, flattened ganglion, about the size of a pin's head (Gray); is surrounded by fat; about 2''' long, in the middle about 0'''.9, and at the anterior and posterior ends about 0'''.78 broad. It lies on the outer and underside of the nervus opticus, about $3\frac{1}{2}'''$ in front of the foramen opticum, and about 8''' to 9''' behind the posterior pole of the sclerotica. Posteriorly, it is

in contact with the inferior ramus of the nervus oculo-motorius
(*s*, Fig. 63), and covers the branch of the same nerve that proceeds to the musculi obliquus and rectus inferior (*b*, Fig. 63);
externally, it is in relation with the musculus rectus externus
at the point where the nervus abducens enters it. This ciliary
(*ophthalmic, lenticular*) ganglion is located at the point of union
of three nerves,—of the ramus naso-ciliaris of the first branch
of the nervus trigeminis (*h*, Fig. 63), one branch of the oculo-
motorius (*d*), and one branch of the nervus sympathicus (*r*).
The branch passing from the trigeminus to the ciliary nerves

Fig. 63.

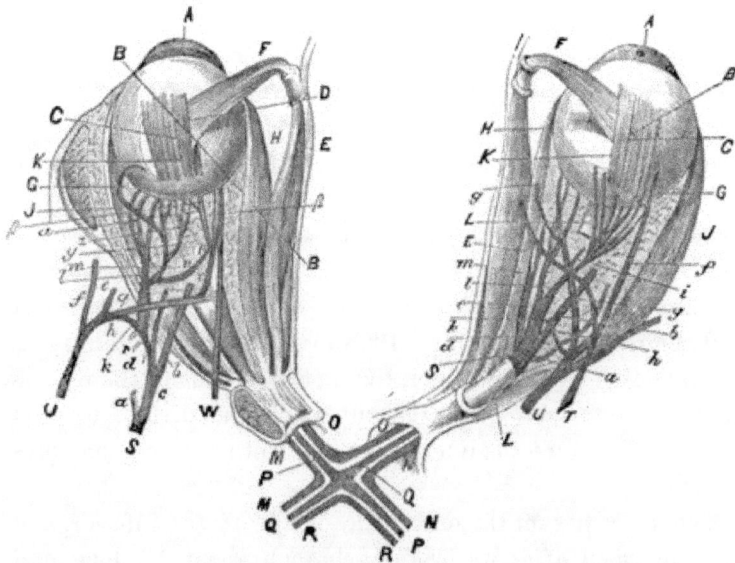

proceeds from the nervus naso-ciliaris (*g*) of the ramus primus
trigemini (*U*), and the ramus secundus of this nerve also sends a
fibre from the ganglion spheno-palatinum to the ganglion ciliare.
It also sends off, after having passed through the fissura orbitalis superior, two branches, the radix longa ganglii ciliaris (*h*)
and the nervus ciliares longus (*i*), the latter of which sometimes is double. The former is 3‴ to 4‴ long, and is found

on the inner side of the nervus naso-ciliaris (*g*), and is covered
by the origin of the levator and rectus superior muscles (*K*),
and also by the ramus superior nervi oculo-motorii, and enters
the outer and posterior corner of the ganglion ciliare (*k*). The
nervus ciliaris longus, at the point where the nervus naso-
ciliaris (*g*) runs above the nervus opticus (*L*), is given off
about 4′″ higher up than the radix brevis from the naso-
ciliaris, turns outward, and lies here on the outer surface of
the nervus opticus. After running along about 3½′″, a few
filaments coming from the ciliary ganglion unite with it.

From hence the nervus naso-ciliaris proceeds to the upper
border of the musculus rectus internus (*H*), and passes out of
the hollow cone formed by the muscles, runs along between
the above muscle and the musculus obliquus inferior (*E*) along
the inner wall of the orbit, and sends off through the foramen
ethmoidalis anterius the nervus ethmoidalis seu nasalis anterior,
and ends as the nervus infra-trochlearis at the inner angle of
the eye, where it divides into the ramulus ad saccum lachry-
malum, rami conjunctivæ palpebrales and nasales.

The nervus oculo-motorius (*S*) gives off in the muscular cone
the ramus superior (*a*), passing to the levator palpebræ superior
and the rectus superior, and 1′″.8 further forward divides itself into two twigs, of which the inner proceeds to the musculus rectus internus (*c*), and the outer (*b*) again divides to send
a branch to the rectus inferior, and the other to the obliquus
inferior.

The latter lies on the outer side of the rectus inferior, and
gives off, soon after its origin, a branch about ½′″ long and
⅓′″ in thickness, the radix brevis ganglii ciliaris (*d*), which,
after passing over the nervus abducens, enters the ciliary
ganglion.

Here, to the outer side, and partly over the short root, and
also over the ramus inferior oculo-motorii, are the arteria ophthalmica and the nervus opticus.

Sometimes twigs from the oculo-motorius are sent to the
rectus externus and the obliquus superior, and frequently

there is a communication between the *ramus superior oculo-motorii* and the *nervus naso-ciliaris*.

There proceeds from the nervus sympathicus, out of the plexus caroticus, in the sinus cavernosus, a thin filament, which passes through the middle division of the *fissura orbitalis superior*, called *radix media* or *trophica ganglii ciliaris* (r), and unites with the ciliary ganglion.

There are certain anomalous connections with the *ganglion ciliare*, such as the *radix inferior longa*, which comes from the *nervus naso-ciliares*, beyond the optic nerve, or from a free ciliary nerve, which lies beneath the *nervus opticus*, and forms a nerve ring, by connecting with that part of the *nervus naso-ciliares* which rests on the optic nerve, and which passes through the *nervus opticus*. Again, a root of the *nervus lacrymalis* passes to the *radix longa*. Another anomaly has been noticed: the origin of the *radix longa* from the *nervus abducens*.

Out of the *ganglion ciliare* there proceed two nerve fasciculi, an inner thicker (m), and an outer and thinner (l) (Budge, Pilz). The former again divides into an inner and outer fascicule. From the inner part of the outer fascicule a nerve proceeds (n), which runs beneath the optic nerve, and connects itself with the *nervus ciliaris longus internus*, from the *naso-ciliaris* (x), at which point (p) Faesebeck asserts to have found a *ganglion ciliare internum*.

This has, however, been denied by Hyrtl and Budge, and the sum of the matter is, that both nerves are in contact, and from this union a twig proceeds beneath the nervus opticus outward, which penetrates the sclerotica at some distance from the optic nerve; and two twigs, which enter the sclerotica behind the optic nerve, and which have between them the *arteria ciliaris longa interna* (W). The second nerve from the inner part of the inner fascicule (y), lies close to the outer side of the nervus opticus, and divides into a number of branches, part of which proceed to the bloodvessels, and others are lost in the fatty tissue. One of these branches (z) passes beneath the nervus opticus, and unites with the branch passing outward that was first described. From the outer part of the inner fascicule 4 to 5 twigs originate, all of which penetrate the sclerotica close to the optic nerve, with the exception of one twig, which runs some distance below the *rectus superior* to the *rectus internus*, and sends off vascular twigs, which also enter the sclerotica near the entrance of the optic nerve.

The outer fascicule divides into three filaments, one of which, far inward, ascends over the optic nerve, whilst the other two pierce the sclerotica more outwardly. They send off vascular twigs, one of which, it is said (Hirzel and Tiedemann), penetrates the nervus opticus and passes to the arteria centralis retinæ. Besides the nerves forming the *ganglion ciliare*, the sixth pair of cerebral nerves enter the cavity of the eye-muscle cone. This *nervus abducens* enters the rectus externus at its origin in the tendinous ring, and passes to the inner surface of the muscle.

Through the upper division of the *fissura orbitalis superior*, formed by the tendinous ring, three nerves enter the orbit,— the *nervus trochlearis*, or fourth pair of cerebral nerves, and two branches of the first ramus of the *nervus trigemini*, the *nervus lacrymalis* and the *nervus frontalis*.

The *nervus trochlearis* (Fig. 63, *D*), at the posterior part of the hollow muscle cone, lies on the outer side of the nervus frontalis, but runs over it in its course forward, passes over the upper surface of the levator palpebræ muscle, to the upper surface of the musculus obliquus superior.

The *nervus lacrymalis* (*e*) passes into the orbit through the outer angle of the fissura orbitalis superior, further outward than the nervus trochlearis and frontalis, through a peculiar small canal in the upper division of this fissure, and runs along the upper border of the *musculus rectus externus*, on the wall of the orbit, to the outer eye-angle, where it terminates in the rami conjunctivales palpebrales and cutanei externi. In its course it anastomoses with the nervus subcutaneus malæ,

the second branch of the trigeminus, and also sends branches
to the lachrymal gland.

The *nervus frontalis* (Fig. 63, *f*) runs along between the roof
of the orbit and the upper surface of the musculus levator palpebræ superioris to the foramen supraorbitale, to pass into the
forehead; as it emerges from the forehead, it gives off a small
branch, the nervus supra-trochlearis, which passes through
the suspensory band of the trochlea to the eyelid; it forms an
anastomosis with the *nervus infra-trochlearis.*

It will be observed that all the nerves of the orbit described
pass through the fissura orbitalis superior, with the exception
of the optic nerve. Another nerve, the *nervus subcutaneus
malæ*, which springs from the *nervus infra-orbitalis*, a branch
from the second ramus of the *trigeminus*, passes through the
fissura spheno-maxillaris seu orbitalis inferior into the orbit.
After a short course, during which it forms anastomoses with
the *nervus lacrymalis*, it passes out again through the *foramen
zygomaticum orbitale*. The main branch, the *nervus infra-orbitalis*, passes through the *canalis infra-orbitalis*, and ends in the
integument as the *rami palpebralis, nasales laterales*, and *labialis
superiores*.

As regards the deep origin of these nerves: The *nervus oculo-motorius* proceeds from the inner side of the pedunculi cerebri,
where some of its fibres are connected with the substantia
nigra and the tegumentum, the fibres passing backward and
inward to the floor of the aquæductus Sylvii, where they are
lost in the gray substance.

The *nervus trochlearis* and the *nervus abducens* have an origin
like the oculo-motorius and the motor roots and muscular
branches of the spinal nerves. The *nervus trochlearis* originates from the laqueus, immediately back of the tubercula
quadrigemina, and immediately through these, from the anterior column of the spinal cord, which latter receives its
fibres from the pyramid of the elongated marrow at the
posterior border of the pons Varolii.

Although the nervus oculo-motorius, the nervus trochlearis,

and the nervus abducens, are some distance apart after their passage from the brain, they yet have a common direction in their course from their origin at the posterior part of the base of the brain to the fissura orbitalis superior of the orbit.

The *nervus oculo-motorius* passes into the orbit uppermost, and immediately beside the processus clinoideus posterior of its side, the nervus abducens undermost on the upper side of the clivus, and the nervus trochlearis on the outermost side. In their further course they are surrounded by the *sinus cavernosus* on the outer side of the *carotis;* the nervus oculo-motorius and the nervus abducens, being immediately in contact with the carotis, the former further inward and the latter beneath it, whilst the nervus trochlearis lies to the outer side of the oculo-motorius, which separates it from the carotis.

Still further forward the *trochlearis* remains on the outer side of the nervus oculo-motorius, and only at the tendinous ring of the orbit does it turn more upward. In this whole extent of its course it lies on the outer side of the first branch of the trigeminus, which comes from the ganglion gasseri, which connects itself with the fasciculi of the three motor nerves of the eye, and thus assumes an outer position to the nervus oculo-motorius and the nervus abducens, and an inner position to the nervus trochlearis. During their position in contact with the carotis, the nervus oculo-motorius and the nervus abducens form connections with the plexus caroticus. This connection with the nervus oculo-motorius signifies that the sympathetic root of the ganglion ciliare passes to this nerve, to proceed with it into the orbit, whilst the importance of the connection with the nervus abducens, which at this point is enlarged like a plexus, is not yet known, but which is likely a giving off of fibres to the sympathicus.

As regards the origin of the *nervus trigeminus*, and more particularly of the *ramus ophthalmicus*, Budge supposes that the motor fibres going to the pupil originate above the second cervical nerve, and the others pass in the *corporibus restiformibus* and the *loca cœrulea*. The *trigeminus* does not only form an anastomosis with the oculo-motorius, but also with the abducens, and sometimes with the trochlearis. Through these connections, doubtless, are the nerves supplying the ocular muscles furnished with sensitive fibres.

The fibres of that part of the nervus sympathicus which proceed to the nervi oculo-motorius, abducens, and trochlearis, as well as those given off to the vascular nerves supplying the arteria ophthalmica, which make up the plexus ophthalmicus, come from the plexus caroticus internus, and from that part of the plexus cavernosus within the sinus cavernosus, and from the upper cervical ganglion of the sympathicus. From this plexus some fine connecting fibres proceed, and connect with the *ganglion gasseri*, the *portio major* of the *nervus trigeminus*, and to the *ramus* 1, *nervi trigemini*. The origin of those fibres of the *sympathicus* which go to the iris and the *dilatator pupillæ*, Budge found in the cervical portion. They come on the one side from the spinal marrow and medulla oblongata through the rami communicantes of the second and first thoracic of the eighth and seventh cervical nerves, from a lower centre, of which the essential part is the middle column of the spinal cord; on the other side, from a connecting twig below the *ganglion cervicale supremum*, which connects the latter with the *nervus hypoglossus*, from an upper centre, located near the lower.

Budge names the part of the sympathicus which proceeds only to the iris, the *iris sympathicus*, and says that it contains both sensory and motor fibres, the former running from the iris to the spinal cord, and the latter from the spinal cord to the iris. Hitherto it has not been demonstrated that there is any relation between the ganglia through which the iris sympathicus passes and the nerve fibres which are destined to dilate the pupil.

We are as yet unable to determine with certainty what special portions of the brain preside over the function of sight. In the cerebral hemispheres, which are the seat of consciousness, thought, and reason, the impressions made on the retinæ and on the fifth nerve are elaborated, and decisions made on the form, size and character of objects.

When the brain is diseased or injured, the vision may be perfect, but the power to comprehend that which is seen, to connect things, and to form conclusions, is destroyed. The impressions of light are felt, the eyes move, the fifth nerve remains impressible to mechanical irritation, but the power of collecting the impressions is gone, the eyes no longer move under the influence of a will guided by intelligence, but roll around unconsciously.

The optic nerve-fibres run into the anterior *tubercula quadrigemina*, or *nates*, and these latter are the seat of the luminous perceptions. When these tubercles are removed in animals, blindness ensues, the pupils dilate, and remain motionless. The removal of one of these eminences causes blindness of the eye on the opposite side: the pupil will dilate, and act only in sympathy when the other eye is impressed by light. The retina, however, is the organ on which the luminous impressions are made, and the office of the tubercula quadrigemina is likely purely psychical. The optic nerve-fibres do not exclusively originate in the *nates*, but they also extend into the gray mass of the *thalami optici*, which belongs to that part of the brain presiding over voluntary motion, in which are also lost the radical fibres of the *corpora restiformia*, which, as has been explained, have an intimate union with the *ramus ophthalmicus* of the *nervus trigemini*.

In this manner there is established an intimate union between the impressions of touch and the voluntary movements with the functions of the retina.

In the intimate interlacement of the fibres of the *trigeminus* with the optic fibres in the brain are brought about the power to define that which is seen, to form conceptions of the dimensions of space, etc. The perception of the condition of the accommodation, the power to distinguish it from the impressions made by the contraction of the muscles, is assisted by the *nervus ciliaris longus*, which is double, and which is mostly quite separate from the rest of the ciliary nerves, given off from the *ganglion ciliare*, and proceeds to the *tensor choroidea*.

The *trigeminus* acts through the sensation, to indicate the degree of retinal activity, and acts as a regulator to the quantity and intensity of light the latter can bear, and interferes to protect it by contracting the iris. The *trigeminus* further influences the *nervus facialis* by causing the lids to contract when the light is too bright.

The *trigeminus* is also the active agency that presides over the nutrition of the eyeball, especially of the cornea. It keeps it transparent. It is well known that when the fifth nerve is injured the cornea becomes opaque, the conjunctiva is congested, the ball becomes anæsthetic, the cornea sloughs, and the eye is lost.

The motive power of the eye has its seat in the *pons Varolii* and the *medulla oblongata*. It is well known that injuries of those parts cause complete immobility of the eye with dilatation of the pupil.

The above description of the nerves of the eye, which has been, to a great extent, translated from Pilz, aims to present to the reader a view of their relations and connections with each other. It will aid the student in acquiring a better knowledge of the separate nerves and their ramifications, to give a brief *résumé*.

The *nervus oculo-motorius* is connected with the part of the cerebrum corresponding with the anterior cords of the spinal column. They appear at each edge of the *crus cerebri*, anterior to the *pons Varolii*, and posterior to the *corpora albicantia*, some of its fibres extending into the *locus niger* of the *crura*, and others to the *corpora quadrigemina*. It receives a filament from the sympathetic in the cavernous sinus, and divides into an upper and a lower branch, which enter the orbit between the two heads of the abductor muscle. The smaller or upper branch is distributed on the ocular surface of the rectus superior and the levator palpebræ muscles. It anastomoses by its twigs with the nasal nerve.

The lower and larger branch is distributed on the ocular

surfaces of the rectus internus, the rectus inferior, and the obliquus inferior. It communicates with the *lenticular* ganglion by a short thick branch.

The fourth pair of nerves, the *nervus trochlearis*, is the smallest of the cranial nerves. It is a motor nerve, and is distributed on the ocular surface of the *obliquus superior*.

The *ophthalmic division* of the fifth pair of nerves (trigeminus) has its central connection with the lateral part of the medulla oblongata, continuous with the floor of the fourth ventricle, being a nerve of sensation. It is the uppermost and smallest division of the *casseroid ganglion*. In the cavernous sinus it receives twigs from the sympathetic plexus, and before entering the orbit it divides into the frontal, the lachrymal, and the nasal nerves.

The frontal, the largest of the three branches, divides into an outer larger branch, the supra-orbital, and an inner smaller, the supra-trochlear, which (latter) passes out of the orbit above the pulley of the inferior oblique muscle, where it subdivides into numerous branches for the muscles and integuments. The supra-orbital escapes at the foramen of the same name, to the brows and forehead.

The nasal subdivides into the proper nasal, and the infra-trochlear. It gives off a branch to the lenticular ganglion. The nasal branch is distributed to the Schneiderian membrane of the nose. The infra-trochlear supplies the lachrymal sac, conjunctiva, eyelids, and neighboring skin with sensation.

The *lachrymal* branch enters the lachrymal gland, and sends filaments into the conjunctiva and lids.

The sixth pair of nerves, the *nervus abducens*, is a motor nerve, having its origin from the pyramidal body of the medulla oblongata, close to the pons Varolii, and is distributed exclusively to the ocular surface of the rectus externus.

The *lenticular ganglion*, belonging to the ganglionic system, or organic system of nerves, is about the size of a pin's head, located at the back part of the orbit, and after receiving twigs from the nasal and the motor oculi nerves, sends off from 14

to 18 posterior ciliary nerves to enter the sclerotica about two
lines from the entrance of the optic nerve.

It will be perceived from the distribution of the nerves that
the cerebro-spinal and the sympathetic systems of nerves are
intermixed in the eye. That the iris and the ciliary body are
supplied with both is evident. Graefe says that the force
which presides over active accommodation is derived from the
cerebro-spinal system ; the other, which holds under its control
the circular fibres of the ciliary muscle, is derived from the
ganglionic system. On this last opium and belladonna act
with opposite effects, the former paralyzes, which permits the
pupil to contract, whilst the belladonna excites those nerves,
and dilates the pupil, by contracting the radiary muscular
fibres.

The Eyelids.

The eyelids are two movable curtains, being continuous
with the integument. They are connected with the border of
the orbit and the globe of the eye by processes of fibrous
membrane or fascia. These curtains or folds are composed
externally of integument, on their inner surface of mucous
membrane, and inclosed between these are the tarsal cartilages, glands, hair-bulbs, muscles, bloodvessels, and nerves.
They close the cavity into the orbit, and lie over the anterior
convexity of the globe, being kept in close apposition by the
action of the muscles, and atmospheric pressure. The free border of the lid measures about 1''', and it has an outer sharp
border, on which are found the *ciliæ* or eyelashes, and a posterior
rounded border, on which are the openings of the ducts for
the Meibomian glands (c, Fig. 64).

The elliptical space between the upper and lower lids is the
palpebral fissure (*fissura palpebralis*), at each end of which is
the union between the upper and lower lids, forming the inner
and outer angles of the eye, or canthi, of which the outer is
more acute than the inner, but the latter is prolonged inward
toward the nose for a short distance. According to the obser-

vations of Arlt, on frozen eyes, the free borders of the lids, when
the eyes are closed, are in complete contact, on their inner
border as well as on their outer, and the triangular space

FIG. 64.

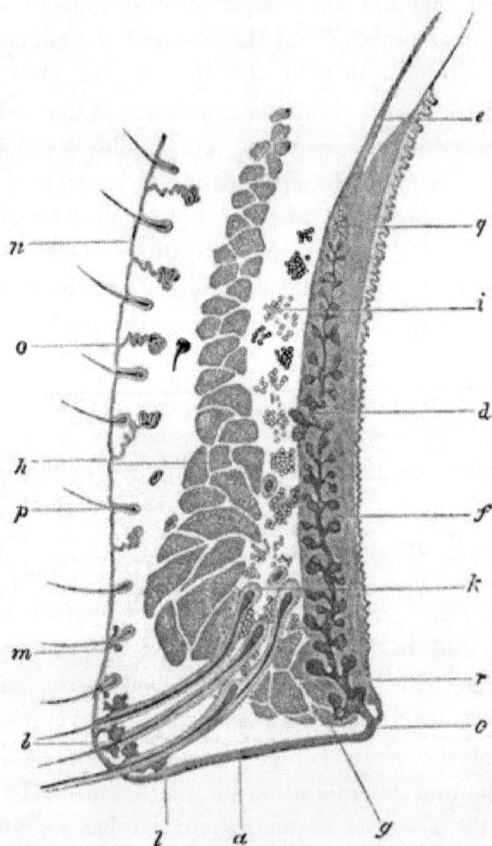

a. Free border of the eyelid. *b*. Outer lip of palpebral border. *c*. Inner lip of the
palpebral border, and also the mouth of the duct of a Meibomian gland. *d*. Tarsal cartilage. *e*. Fascia-like membrane between the cartilage and the orbital border. *f*. Meibomian glands. *g*. Inner portion of orbicularis (musculus tarsalis). *h*. Fibre-bundles of
the orbicular muscle. *i*. Cellular tissue beneath the orbicular muscle. *k*. The bulbs
of the cilia. *l*. Lubricating glands. *m*. Small hair on the skin of the lid. *n*.
The outer skin of the lids. *o*. Sweat-glands. *p*. Very fine hair. *q*. Conjunctiva tarsi.
(*From Stellwag.*)

(*lacus lachrymalis*) between them does not exist, as has been
generally taught in text-books.

The *tarsal cartilages* (*tarsi*) (*d*, Fig. 64) form the firm basis for each lid. In structure they belong to the compactly formed connective tissue, with a certain number of small cartilage-cells bound up with it. They are semilunar in shape, about 0'''.3 to 0'''.4 thick, and are elastic. The upper and larger is nearly one-half inch in breadth at the middle. They have their inner angle more obtuse than the outer. The outer ends project a short distance beyond the canthus. The lower has the same length as the upper, but is much more narrow, thinner, softer, and is of a more fibrous character. The anterior surface of the tarsal cartilages is covered by the *musculus orbicularis palpebrarum*, with which it is connected by a very yielding connective tissue, whilst its posterior surface is firmly attached to the conjunctiva. Toward the orbital border the cartilages become thinner, and terminate in a fascia-like membrane (*e*, Fig. 64), which is connected with the orbital border. This *fascia tarso-orbitalis* is in connection with the tendon of the levator muscle of the lid, which expands into a broad membrane, and is lost in this fascia. The tarsal cartilages are connected with the *margo orbitalis* by a cellulo-fibrous connective tissue mass, which proceeds from its periphery, and in some parts it is like fascia in structure, whilst in other parts it consists simply of loose cellular tissue membrane. Four of these fibrous bands have generally been described, two of which unite the outer and inner ends of the cartilages, named the *ligamentum canthi internum* and *externum*, and two of which proceed from the border of the cartilages (*e*, Fig. 64) to the margin of the orbit, in the form of flat, broad fascia-like membranes, the *ligamentum tarsilatum superius et inferius*. Pilz thinks that the *ligamentum canthi internum* can alone be considered a true ligament. The others are more like cellular tissue, and have no well-defined borders.

The inner palpebral band or ligament (*n*, Fig. 65) is 2½''' to 3''' long, and about 2''' wide, and originates from the periosteum of the frontal process of the superior maxillary bone, runs horizontally outward and is lost (*l*, Fig. 65) in the inner obtuse ends of the tarsal cartilages of the upper and lower lids. It passes over the lachrymal sac, and at the commissure of the lids it divides (*l*, Fig. 65), part accompanying the fibres of the

FIG. 65.

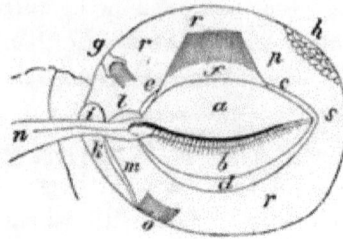

(*From Pilz.*)

orbicular muscle, which proceed from the lachrymal bone to the lachrymal canals, the anterior and outer walls of which it covers; another process passes back to the eyeball, which is continuous with the tunica vaginalis bulbi by a process. It is this process that is adherent to the conjunctiva, and which is an important fact to be noted in connection with the operation for squinting, as pointed out by Liebreich. It supports the semilunar fold (*plica semilunaris*) and the *caruncula lachrymalis* resting on it. Its upper and lower surfaces afford origin for the *portio anterior* of the *musculus orbicularis;* the anterior border is immediately beneath the skin, and the posterior border covers the inner half of the lachrymal sac, with which it is firmly united, and serves to strengthen it.

The *ligamentum canthi externum* (*s*, Fig. 65) is not really a ligament, but consists of a firm fibrous texture, and unites the acute angles of the outer ends of the tarsal cartilages, and with the firmly adherent skin, forms the outer eye-angle, and is blended with the periosteum of the orbital border, and with the process of Tenon's membrane. The cellular tissue membrane that fills up the space above the *ligamenta canthi*, between the tarsal cartilages and the border of the orbit, has been described by authors as the *ligamenta lata*. This cellular tissue in certain points is condensed into a firm structure, which appears tendinous, and throughout its whole extent is loosely connected with the orbicularis muscle, which rests on it, and partly insheathes its fasciculi ; portions of it are attached to the cellular tissue immediately in contact with the skin of the lids.

For a more special description of the cellular tissue membrane of the lids, a division of the parts above and the parts below the ligamenta canthi becomes necessary. In the cellular membrane of the upper division but three points are found where a true fibrous structure is manifest: at the point of insertion of the *musculus levator palpebræ superioris* (*f*, Fig. 65), outward where it covers the lachrymal gland (*h*, Fig. 65), and to which it sends processes (*p*), and over the *trochlea* (*g*), from which a suspensory band is given off to the *bulbus*. In all other portions of this part of the membrane it consists of cellular tissue.

In the lower division of this cellular membrane, only one point is characterized by the fibrous structure, which is a band, and extends from the middle of the lower surface of the *ligamentum canthi internum*, passing obliquely outward and downward (*m*, Fig. 65), to a portion of the rim of the orbit between the lachrymal sac and the origin of the *musculus obliquus inferior*, and covers the outer surface of this muscle (*o*). In all other parts this membrane is composed of cellular tissue. Nearer the tarsal cartilage it disappears in the stroma of the conjunctiva at the lower cul-de-sac (*d*, Fig. 65), and the band of Tenon's membrane given off in that direction.

The *Meibomian glands* are imbedded in the tarsal cartilages, nearer to the posterior than the anterior surface. On everting the lids they can be seen as parallel strings of pearls running in a vertical direction. In structure and in the character of their secretions they are like the sebaceous glands (*f*, Fig. 64). They are in number about 30 in the upper cartilage and 20 to 25 in the lower. They do not extend throughout the whole breadth of the cartilages, not reaching their attached border, and are surrounded on all sides by the cartilaginous substance.

In the main they are regularly vertical in their direction, but sometimes their posterior ends are united, or they bend over laterally and form a curve. Toward the palpebral border the glands become larger. In structure they present numerous, somewhat round, angular *cryptæ aggregatæ*, with a diameter of $\frac{1}{20}'''$ to $\frac{1}{8}'''$, which surround in a horizontal direction a canal of $\frac{1}{12}'''$ to $\frac{1}{4}'''$, and in certain cases $\frac{1}{4}'''$ in diameter. The secretion of the Meibomian glands is like that of the sebaceous glands, and it is said (Kölliker) that the only difference is that the fat does not collect into large drops, but remains in separate particles. At the border of the lids it excretes the eye-butter, the *lema palpebralis*, which is a whitish, rather thickish, fatty substance, intended to prevent the adhesion of the lids.

The tarsal cartilages do not reach to the free border of the lids within $\frac{1}{4}'''$ to $\frac{1}{2}'''$, being covered by ordinary epidermis and mucous membrane (Fig. 64). The ducts open on the free borders of the lids (*e*, Fig. 64), the mouths of which are $0'''.04$ to $0'''.05$ in diameter. The tubes consist of basement membrane, covered by a layer of scaly epithelium. The outer skin of the lids is delicate and thin, easily raised into folds. It has a thickness of about $\frac{1}{6}'''$ to $\frac{1}{8}'''$. Its *tela cellulosa subcutanei* is poor in fat, and its inner surface is connected by a loose cellular tissue to the orbicularis muscle and the palpebral bands above described.

The outer layer of the subcutaneous cellular tissue is connected with the corium by a large number of cords of connective tissue, and is not sharply divided from it. This subcutaneous cellular tissue varies in thickness according to age, sex, and individuality, depending chiefly on the amount of fatty tissue in it. According to Krause, the thickness of the subcutaneous cellular tissue, void of fat, is $\frac{1}{4}'''$. The *corium*, which is an inelastic skin, mostly made up of connective tissue, consists, in most parts, of two layers, especially where it rests on adipose tissue—the *pars reticularis* and *trabecularis*. In the eyelids there is only one layer, the structure of which has the *pars papillaris* with very short papillæ, which in fact are sometimes entirely wanting. The corium is $\frac{1}{5}'''$ to $\frac{1}{8}'''$ only in thickness.

The skin of the eyelids contains sweat-glands (Fig. 64), which extend to the very border, and measure about $\frac{1}{10}'''$ to $\frac{1}{12}'''$, and are not located, as is common, in the pars reticularis, but are in the subcutaneous cellular structure, or in the border thereof. The ducts of the glands have delicate walls, and are without muscles. The skin also possesses many small hairs (Fig. 64), which differ from the ciliæ in not having sebaceous glands at their sides (Kölliker).

Toward the orbital border the skin becomes thicker, and toward the lower border of the orbit it is richly supplied with adipose tissue.

The *orbicularis palpebrarum* muscle (*musculus orbicularis*) is a sphincter muscle, which is expanded beneath the integument of the lids, and around the circumference of the orbit, extending some distance beyond the border of the latter (see Figs. 66 and 67). We are indebted to Professor Arlt, of Vienna, for a clear description of this muscle, and what follows is mostly drawn from him. (*Archiv*, 1, x, 2 ; *Compte rendu*, 1862.) Beneath this muscle are the two tarsal cartilages, and the facial surface of the bones all around the orbit, and the membrane between the cartilages and the orbital border, the *fascia tarso-orbitalis*.

Below the internal palpebral ligament a small portion of its fibres rest on the anterior surface of the lachrymal sac. Toward the periphery of this muscle there are located, in places, other muscles between it and the bones, and in other localities masses of fatty tissue. Over the opening of the orbit there is not much fatty tissue in contact with the muscles. Behind the palpebral ligaments is the fatty tissue of the orbit. Backward from the outer commissure of the lids the outer palpebral ligament is attached, composed of connective tissue of firm texture, and rich in elastic fibres, which is there connected with the orbital surface of the malar bone, and with the tunica vaginalis bulbi, which retains the outer angle of the eye a considerable distance

from the orbital border. Immediately above this band is located
the lower part of the lachrymal gland. At the bone it projects
2′″ within the orbital cavity.

FIG. 66.

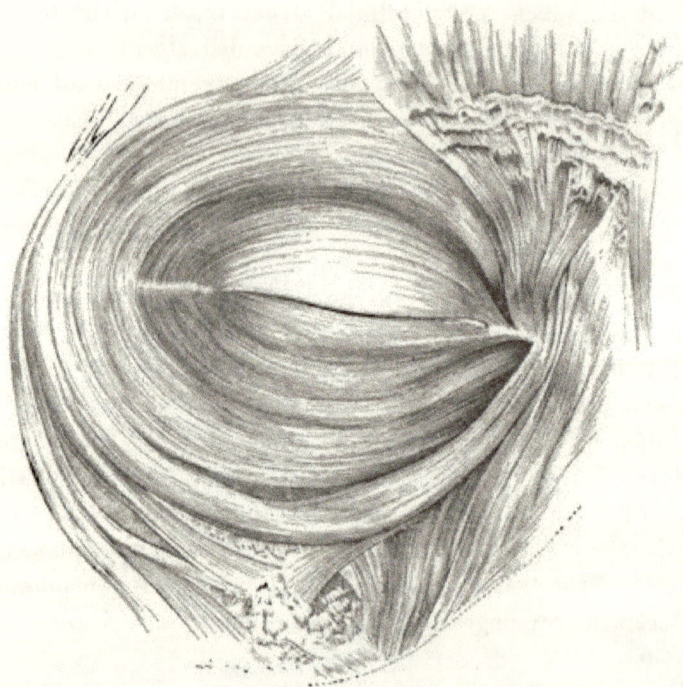

The *fixed points* of the orbicularis are mostly situated at the
nasal side, some also on the temporal side; a portion of the
fibres along the periphery have variable points of attachment,
being in a measure dependent on the contractions of other
muscles.

The attachments of the muscle at the nasal side are at the
crest of the lachrymal bone, at the inner lid-band (tendo-oculi),
at the orbital border, at the facial surface of the superior
maxillary, and the frontal bones. As the *ligamentum canthi
internum* is so intimately mixed with an intelligible description
of the orbicularis, a further description of it is required here.

The *inner lid-band* (tendo-oculi) is closely connected with the thin skin covering it, and is visible through it. By pulling

Fig. 67.

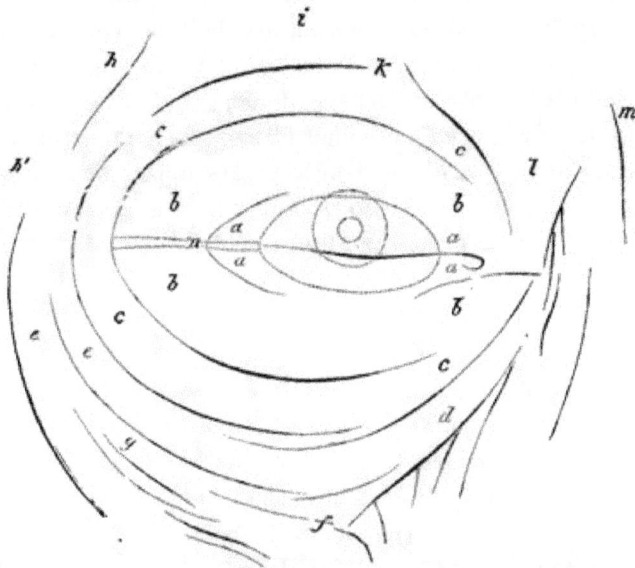

a, a, a, a. Lachrymal part of the orbicularis muscle (Horner's muscle) nearest the *ligamentum canthi internum,* and partly covered by the fibres of the *lid-band* portion of the muscle.

b, b, b, b. Extent of the *lid-band* portion of the muscle, nearest the *lid-band* (*tendo-oculi*), and above it, part of it rising out of the deeper parts beneath.

c, c, c, c. Extent of the orbital portion, the origin of the fibres in a great measure not being visible.

d. Descending branch of the peripheral portion, sending fasciculi to the skin at *f,* and also some downward toward the angle of the mouth.

e, e. Ascending part of the peripheral portion, sending almost constantly one or more bundles, as seen at *g,* to the angle of the mouth.

h. Bundles extending to the aponeurosis of the muscles of the forehead, and sometimes, as seen at *h,* to the temporal aponeurosis.

i. Passage of the peripheral fibres to the muscles of the forehead.

k. Points where the fibres from the third and fourth divisions of the muscles pass to and beneath the eyebrows.

l. Triangular layer of the peripheral fibres, between the lid-band and the inner half of the eyebrows.

m. A slender bundle of fibres, extending from the dorsum of the nose to the eyebrows and integument of the forehead.

The borders of the tarsal cartilages, the position of the cornea in a state of rest, and the longitudinal line where the fibres of the first and second parts have a peculiar relation, are marked with *dotted lines.*

n. Shows the point where the outer lid-band touches the bone.

(*From Arlt.*)

the outer commissure of the lids outward, and a little upward, with the thumb, its whole length may be seen. Its free border is a little more than 3''' long, and passes outward across the lachrymal sac. Its smaller, inner half rests on the superior maxillary bone, and its outer larger half on the lachrymal sac. Its under surface, which forms an acute angle with the anterior surface of the lachrymal sac, gives off, on its whole length, closely packed muscular fibres, which are attached to the anterior surface of the tear sac. Its upper surface, which immediately passes over into the thick fibrous covering of the tear sac, gives off muscular fibres to the lids only in its outer half.

Those given off from the inner half of its surface do not belong to this muscle, as will be explained hereafter. The inner end of the ligamentum canthi internum is gradually lost in the periosteum of the superior maxillary bone. Outward this lid-band is divided into two horns, which may also be seen through the skin, and which form the triangular space between the lids which incloses the caruncula lachrymalis. These processes of the lid-band serve for the attachment of numerous muscular fibres, and are gradually lost about half way in their course to the puncta lachrymalia. Behind these processes the lachrymal canals sink deep to perforate the muscle coming from the crest of the lachrymal bone, so as to reach the outer wall of the lachrymal sac, into which they open. The orbicularis completely surrounds the orbit and eyelids, its peripheral bundles to a certain extent forming isolated parts. The muscular bundles of its central half originate from the crest of the lachrymal bone, and from the inner lid-band are quite thin and pale, especially in the region of the upper lid.

They are arranged singly, side by side, with the exception of the region near the border of the lids, where they rest on each other. At the temporal side, where the upper and lower halves meet, there is a peculiar arrangement, soon to be noticed. Above and below the middle of the palpebral fissure, the periphery of this portion of the muscle about corresponds with the border of the orbit; outward from the fissura palpebralis these muscular fibres extend 7''' to 8''' from the outer commissure on to the malar bone.

The bundles of fibres of the peripheral half originate from that part of the superior maxillary bone which borders on the lachrymal groove, on to the region of the infra-orbital canal, and form a line less marked on the frontal bone, which rises in the region of the crest of the lachrymal bone, on to the *incisura supra-orbitalis*, partly from the facial surface of the bones along the described line of insertion, and from the lachrymal sac. The different parts of this muscle will be described according to the origin of their muscular bundles.

1. The *portio lachrymalis* (Horner's muscle, which Arlt considers a part of the orbicularis) arises by a short tendon, about 3''' in breadth from the upper third of the crest of the lachrymal bone. From its origin to its division it is flat and oblong in shape (see Fig. 68).

The somewhat even surface looking toward the globe of the eye, receives from the periorbita a covering of firm aponeurosis, which, toward the bifurcation, is in connection with the palpebral ligament as well as with the tunica vaginalis bulbi. This connection is not immediate, but is formed by connective tissue, rich in elastic fibres, to which the caruncula lachrymalis is attached in front. Shortly before the division the aponeurosis and the connective tissue are so firmly blended, that in their course to the cartilages it is difficult to separate them. The opposite surface is free only nearest the upper and lower edge; in the middle part it is in contact with the upper portion of the lachrymal sac, and further forward it is attached to the outer end of the lid-band.

From the bifurcation to the inner extremities of the tarsal cartilages the muscle forms two cords, almost round, about the size of a raven's quill, which envelops the lachrymal canals. Laterally, they are uninterruptedly connected with the radiating muscular fibres of the orbicular muscle, emanating from the lid-band, and near the cartilages expand to be inserted into their inner extremities. But few of these fibres next to the

Fig. 68.

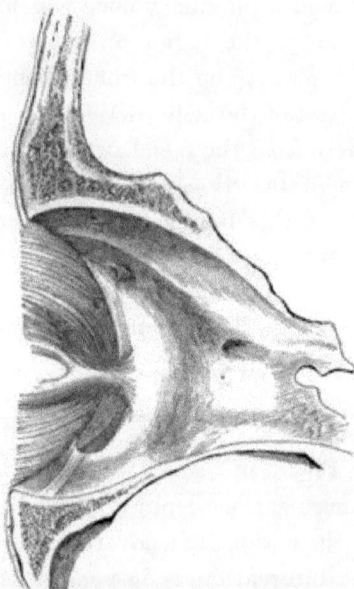

The orbicularis, seen on its posterior surface. The inner surface of the orbit seen (nasal side). In the back part the foramen opticum, and further forward the ethmoidal fissure; still further forward the trochlea is seen, above which is the nervus supraorbitalis. Near the lower border is seen a piece of the musculus obliquus. The lachrymal sac is also seen, partly covered by Horner's muscle. (*From Arlt.*)

free border of the cartilages, and therefore beneath the hair-bulbs of the cilia, end in the cartilage without passing beyond the outer commissure. The breadth of this linear band of fibres is $\frac{1}{2}'''$ to $\frac{3}{4}'''$. The remainder of the fibres pass over the loose fascia given off from the palpebral membrane, and are therefore not immediately in contact with the cartilages, and only near the hair-bulbs are they in connection with it. Toward the outer angle of the eye some fibres seem to be inserted into the border of the cartilage. Those situated nearest the border of the lid run parallel with it, and cover the hair-bulbs; those further off form large curves, with the convexity toward the orbital border of the cartilages, the cornu growing larger as they approach nearer the orbital border.

On the temporal side the muscular fibre-bundles of the upper lid meet those of the lower lid at somewhat sharp angles, which grow less acute the further away from the canthus they meet. Those further from the outer canthus reach to or beyond the bony rim of the orbit.

2. The *portio anterior, vel portio ligamentum palpebrale internum*, of the orbicular muscle, is in its origin even much stronger in its lower half than in its upper half. The fibres originate the whole length of the lid-band, out of the sharp angle between it and the lachrymal sac, and are so firmly connected with the fibrous covering of the latter that they even seem to originate from it. Anteriorly they are covered by a

Fig. 69.

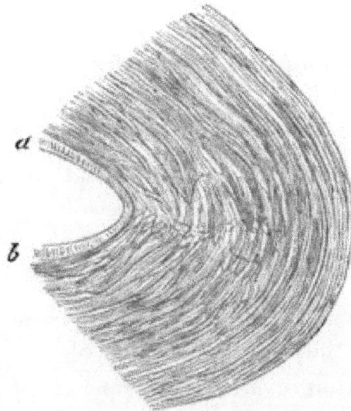

Fig. 69 represents the outer part of the orbicularis. *a.* The upper cartilage. *b.* The lower cartilage. The mode of union between the fibres between the lower and upper halves of the muscle is seen. (*From Arlt.*)

short, unyielding connective tissue, and through this is the thin skin devoid of adipose tissue; backward they are connected by sheaths to the bones toward the nose. The angle which they form with the longitudinal direction of the lid-band can readily be estimated when it is known that the fibres originating at or near the outer end of the lid-band nearly cover those bundles coming from the portio lachrymalis. Further in, they pass by the side of them, and those originating from the inner or nasal end of the band follow about the direction of the orbit, and at the middle of the palpebral fissure they cover it, and from this point outward they pass over the facial surface of the malar bone to reach there when the lids are closed, a point 7''' to 8''' outward from the outer canthus.

In this portion, as in the former, the muscular fibre-bundles of the lower lid are much larger than those of the upper lid.

The numerous, but very delicate, muscular fibre-bundles of this part in the upper lid originate from the outer end of the lid-band, from the point where the fibrous covering of the lachrymal sac is inserted into the frontal bone, and form curves from the convex border of the cartilage on one side to the bony border of the orbit, first running upward and then outward.

3. The *portio orbitalis* has its bundles of fibres different from the portions just described, being more powerful and darker in color, and on the temporal side they pass from the upper half into the lower without any interruption.

The fibres originate from the superior maxillary and from the frontal bones. On the superior maxillary bone the line of insertion is beneath the lid-band, along the border which helps to form the lachrymal fossa above, and below divides the orbital surface from the anterior surface of the upper jaw-bone. The line of insertion in the frontal bone arises by a surface not well marked, beginning in the direction of the crest of the lachrymal bone, and ends at the incisura supra-orbitalis. The upper end of the lower curved line of insertion lies further forward (almost the breadth of the lachrymal groove) than the lower end of the upper curved line. The lid-band is inserted between the two, the fibres from which take between them the fibres coming from behind from the portio lachrymalis.

The fixed points for the upper half of the orbicularis are therefore deeper, or located further back.

The fibres of the lower half of the muscle run obliquely downward and outward, and at the foramen infra-orbitalis, and between it and the orbital border, they are partly covered by the more pale and thin fibres of the portio ligamentum palpebrale internum (or lid-band portion). The bundles of one part lie side by side, and also on each other.

In the inner half the bundles of fibres are closely packed. In the outer half they are more broad and flat, and are placed side by side.

The fibres of the upper half, which nearest their origin are partly covered by the lid-band portion, come in contact with the incisura supra-orbitalis, up to which they constantly increase behind, so as beyond it to run along the projecting border of the orbit, and on the facial surface of the malar bone immediately pass over into the fibres of the lower half. At this point the innermost fibres are 4''' to 5''' distant from the orbital border, and are separated from the subjacent bones by some fatty tissue, which is not the case in the second portion of the muscle.

4. The parts of the muscle hitherto described form an uninterrupted plate, with the exception of the palpebral fissure. A considerable number of fibre-bundles are given off, called the peripheral or accessory part of the muscle.

The bundles are thick, and dark red, and are not interrupted at the temporal side.

A strong bundle of muscle-fibres are given off from the bone at the inner end of the lid-band, which passes almost in a straight line toward the canine fossa. For some distance upward fibres are given off from the bone along the posterior surface of the muscle, and therefore it becomes thicker from before backward than from side to side.

On its outer border there is a wrinkle or furrow, which is bounded on the other side by the portio anterior and the portio

orbitalis, as the tense connective tissue which covers the lid-
band portion on the lachrymal sac ascends in the furrow, and
by lateral extensions is fixed to the bone.

The *vena* and *arteria angularis*, which come from the canine
fossa, lie in this furrow. In front of the infra-orbital foramen
the band of muscle attached to the skin of the cheek by numerous bands of connective tissue, turns outward, and then bends
suddenly upward and outward toward the malar bone, where
it is attached to the orbital portion. In front of the canine
fossa some thin bundles are attached to the skin. Sometimes
a bundle is given off to the angle of the mouth. The ascending
branch, after passing a line formed by the extension of the
palpebral fissure to the temple, passes into pencils of extending
fibres, which are lost in the aponeurosis of the frontal muscle.
Downward some of the bundles are lost in the temporal aponeurosis. The rest of the fibres pass above the projecting orbital
border inward, and pass to the *musculus corrugata superciliorum*.
Above the lid-band a flat, muscular layer arises, about 3''' to
4''' broad at its origin. It arises from the bone above the
tear sac. It has a dark color, and ascends toward the superciliary arch, and becomes broader as it approaches it. From
the lid-band it is covered by a fascia of connective tissue,
which attaches it to the skin, and by its lateral extensions it
is fixed to the bones. The fibres of this muscle are lost in the
inner half of the eyebrow. This flat muscle forms a triangle,
its base being in the eyebrow, and its apex at the upper surface of the lid-band. This flat muscle may be named the
abductor of the eyebrows.

Beneath it are found first the beginning of the orbital portion, more toward the nose and upward, the corrugator super-
cilii, which is covered with a somewhat thick aponeurosis, the
fibres of which, below the superciliary arch, run obliquely
outward more than upward, and has connections with the
beginning of the frontal muscle, as well as with the peripheral
fibres of the orbicularis coming from the temple.

The orbicularis extends an inch outward and upward. Of

this two-thirds are included in the lachrymal and lid-band portions. Only these portions present the peculiarities to be named. If a line be extended in the direction of the palpebral fissure of the closed lids, outward from the outer canthus 7''' to 8''' on the malar bone, we shall find that from the angle of the lids to the end of this line the cuticle is thin, without a fatty cushion, and is fastened to the muscle by a connective tissue so short that it is difficult to separate them. Below this line the muscle bundles are thick, and above it they are unusually thin. The fibres of the upper and lower halves approach this line seemingly as if they approached each other at acute angles; but, in front at least, no angles are seen, but bows, or curves of fibres that pass over from one half to the other. The nearer these bows (of fibres) approach the palpebral fissure, the more they curve, and the further they are from it the more flat they are. Within the extent of this line the muscle is also more firmly connected to the firm connective tissue than in surrounding regions. From 7''' to 8''' outward from the external canthus, there is some fatty tissue lying on the muscular fibres, as well as beneath them. Near the line named it is observed that the muscular bundles split, yet it is seen that in its whole extent fibres pass uninterruptedly from the lower to the upper half; but this takes place more freely toward the temple than further inward. Still, it is observed that many fibres here cease by insertion into the firm connective tissue beneath.

The peripheral muscular fibre-bundles have nothing to do with the closing of the eyelids, but are called into action during winking, laughing, or crying. It will be perceived from the above, that Arlt gives more points of attachment for this muscle than anatomists have done heretofore. These he calls the external points of fixed attachment, such as a bundle going to the angle of the mouth, another to the temple, and another to the epicranial aponeurosis, with many of its fibres uniting with the frontal and superciliary muscles.

The external angle of the eye is placed 2''' to 3''' higher

than the internal fixed point of the lids. During blinking, or
gentle closure of the lids, the extremity of the external angle
of the lids descends to a horizontal line with the point of reunion with the internal angle, which is situated near the middle
of the palpebral fissure. During blinking, or closure of the lids,
the upper lid moves in a vertical direction; the middle of the
lid descends until the internal and external fixed points of the
lids are found in a straight line. The lower lid, on the contrary, moves laterally. Whilst the external angle descends,
the internal part of the lid, especially the lachrymal point, is
drawn inward, backward, and upward. The middle of the
border of the lower lid is not displaced upward, and only undergoes a slight inward movement. The two lids, during
blinking, or closure of the lids, make also a slight inward
movement. In the blinking, to save the eye from too great
a light, the peripheral fascicule narrows the arch which it
forms, and the internal portion raises the lower lid, and at
the same time carries it inward. By this action the palpebral fissure is narrowed from the side of the nose.

During laughing the skin is wrinkled toward a line which
runs between the external fixed point of the muscle (7''' to 8'''
outward from the external canthus, described above) and the
external angle of the lids. The narrowing of the palpebral
fissure in this instance takes place by the lower lid being
raised toward the external angle of the lids. In crying, the
lower lid is wholly raised, but the upper lid also descends. In
these movements, as well as those that take place in laughing,
the palpebral fissure is somewhat shortened.

The action of the lachrymal portion of this muscle (Horner's
muscle) will be given in connection with the lachrymal apparatus.

The *ciliæ* or *eyelashes* are located on the outer edge of the
free border of the lids, and very seldom consist of a single
row, but mostly two or three rows exist, one close behind the
other. In the upper they are more abundant and larger, and
are curved downward and outward; in the lower lid they are

less numerous, smaller, and are curved upward and outward. They are short thin hairs, and differ from other hair in being thickest at the lower part of the shaft, from which point they taper to a point both ways.

In their general structure the cilia are similar to other hair, and they do not demand a full description. The hair-follicles are 1½''' in length, and frequently reach into the subcutaneous cellular tissue, and are attached between the ciliary border of the tarsi, and the innermost fibres of the portio lacrymalis of the orbicular muscle.

Close to the hair-follicles some sebaceous glands are located, generally two in number, which open into it. These glands measure about 0'''.06 to 0'''.24 (*l*, Fig. 64).

Near the cilia are numerous small hairs (*m*, Fig. 64), the follicles of which are also furnished with beautifully developed sebaceous glands. The eyelashes are subject to a continual change. They reach their normal length in about 150 days (Stellwag), when their bulb is loosened (*l*, Fig. 64), whilst on the papilla a new hair is developed, which carries the old in front with it, to be removed by rubbing or washing.

The arteries of the integument of the lids, the orbicular muscle, and the tarsal cartilages (in which each Meibomian gland has its own vascular apparatus) have their origin from the arteria ophthalmica (*e*, Fig. 70), which, after giving off the arteria dorsalis nasi (*f*, Fig. 70), and the arteria frontalis (*g*), divides at the inner angle of the eye into two branches, the arteria palpebralis superior and the arteria palpebralis inferior. These branches, after sending twigs to the lachrymal sac, the caruncula, and the conjunctiva palpebrarum, penetrate each one its lid, and run between the tarsi and the sphincter muscle, at least a line from the free border of the lid, outward, and form numerous anastomoses with the vessels around the eyelids.

The *palpebralis superior* (*r*, Fig. 70) anastomoses with the arteria temporalis superficialis (which is a branch of the arteria temporalis from the carotis externa) (*c*, Fig. 70), and with the

arteria supra-orbitaria (a branch of the arteria ophthalmica).
The inferior palpebral artery anastomoses with the arteria
transversa faciei (b), a branch of the arteria temporalis, with
the arteria palpebralis externa (i) from the lachrymal artery,
which is a branch of the ophthalmic, and with the arteria
infra-orbitalis (a), a branch of the internal maxillary artery.
The largest of the branches of the palpebral arteries is the
ramus tarseus seu marginalis, which runs along near the
margin of the lid, still in front of the tarsus, and forms, in
the upper lid, with a branch of the arteria temporalis super-
ficiales anterior, and in the lower lid, with a branch of the
arteria lacrymalis and transversa faciei, a vascular bow parallel with the palpebral fissure, called the arcus tarsus superior
and inferior (k, k, Fig. 70).

Fig. 70.

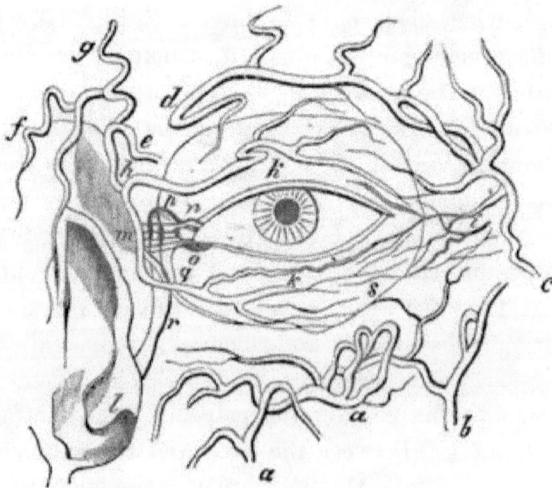

a. Infra-orbital artery.   b. Transverse facial artery.   (a). Their anastomoses.   c.
Anterior superficial temporal artery.   d. Supra-orbital artery.   e. Ophthalmic artery.
f. Arteria dorsalis narium.   g. Frontal artery.   h. Internal palpebral artery.   k. Arcus
palpebralis superior et inferior.   i. External palpebral arteries from the lachrymal artery.   l. Lower muscle of the nose.   m. Ligamentum palpebralis internum.   n, o. Upper
and lower lachrymal canals.   p, q. Orbital portion of the lacrymal sac.   p. Fundus of
the sac.   (From Pilz.)

The veins of the eyelids and their muscles are collected into the superior and inferior palpebral veins; the former pass into the anterior facial and middle temporal veins, and the latter into the anterior facial veins. The lymphatic vessels of the lids proceed to the superficial facial and sub-maxillary lymphatic glands.

The nerves of the skin of the eyelids originate from the nervus trigeminus; those in the upper lid from the nervus supra-orbitalis (from the frontal nerve of the first branch); and those of the lower lid from the nervus infra-orbitalis of the second branch. Besides these, there are passing to the integument of the lids, palpebral branches from the nervus supra-trochlearis, a branch of the frontal nerve, from the nervus infra-trochlearis, a branch of the ramus naso-ciliaris, from the frontal nerve itself and the nervus lacrymalis, both from the first branch of the trigeminus.

The *conjunctiva* (*Bindehaut* in German), begins at the free border of the eyelids as the immediate continuation of the outer integument, covers the posterior surface of the lids, then passes over to the eyeball to cover the anterior part of the sclerotica, and the whole of the cornea. It is thus divided into four divisions: 1, the *conjunctiva palpebralis*, which covers the posterior surface of the lids ½''' beyond the orbital border of the tarsal cartilages; 2, the superior and inferior reflected portions, called the superior and inferior palpebral folds; 3, the *conjunctiva sclerotica*; 4, the *conjunctiva cornea*. This division of the conjunctiva is not wholly artificial, but is demanded by morphological conditions.

The conjunctiva is a mucous membrane, and possesses an epithelium, immediately beneath which is found the most solid part, the papillary body. By the papillary body is meant the most solid part of this membrane, not necessarily possessing papillæ. We may then speak of the papillary body of the ocular conjunctiva, whilst this membrane does not possess any papillæ (Wecker). The mode of union between the papillary body and the deeper layers, varies according to the different parts. In the middle of the tarsal cartilages the union is effected by a very thin, slightly extensible, cellular tissue; in the reflected portion (cul-de-sac) the conjunctiva is united to the eyeball by a cellular tissue with large meshes, which is quite loose and extensible, so as to permit considerable sliding of the mucous membrane over the sclerotica. On the sclerotica, the bundles of connective tissue which spring from the deep parts, unite themselves with the papillary body, become shorter, and in consequence of their rigidity resemble the cellular tissue forming the general covering of the sclerotica. The papillary body diminishes more and more on the sclerotica up to the corneal border, with the exception of the raised part, which follows the superior and inferior borders of this membrane, and which is named the *limbus conjunctiva*, and only a very thin layer of connective tissue remains, which is lost in the corneal substance.

The *conjunctiva palpebralis* is a reddish membrane of a thickness of $0'''.12$ to $0'''.16$, the cellular layer having a thickness of $0'''.08$ to $0'''.11$, and with a lamellated epithelial covering of $0'''.04$ in thickness. The cylindrical epithelium prevails here under the form of a layer of numerous small cells, which inclose a large nucleus, located near the wall, of a granulated appearance, due to the presence of excessively fine small molecules. The deep layer of cells are more elongated, those of the superior layer being irregularly polygonal. It is here a true mucous membrane, being connected to the tarsi by a short, firm cellular tissue, void of fat. On the free, or inner surface, it is only covered by an epithelium, always moist and slippery, covered with mucus. It possesses a true *textus papillaris*, which gives it the appearance of delicate velvet. This is caused by numerous round projections, composed of bundles of the finest vascular loops, the termination of nerves and a fine cellular tissue, but no lymphatics.

These papillæ, similar to those of the cuticle, are found only in this division of the conjunctiva. They are small and cylindrical, but become larger and more wart-like toward the cul-de-sac, where they are found $\frac{1}{2}'''$ in length. Whilst the papillæ cease with the cul-de-sac, the papillary body by no means ceases there. It is formed by a layer of solid cellular tissue, which is lost little by little, in the sub-conjunctival cellular tissue, varying according to the parts of the conjunctiva where it is found. In the middle part of the tarsi the papillæ are small; toward the posterior border they are more projecting. In the conjunctiva of the cul-de-sac they have a large base, but are less prominent. They vary in size according to the age of the individual, and according to the different parts of the conjunctiva. They may have a height of $0'''.1$.

The union of the inferior epithelial cells with the papillary body is made in such a manner that the surface of the latter is not smooth, and is never covered by a basement-membrane; but the tough fibres of the cellular tissue of the papillaries terminate at the surface of the latter by free and slightly projecting extremities. Even in the conjunctiva bulbi, where the papillæ are wanting, the fibrils of the cellular tissue also terminate by free extremities, which appear as small teeth attached to the exterior surface on perpendicular section. There is no direct union between the fibrils of the cellular tissue with the most inferior epithelial cells.

The two things to be observed as peculiar to this division of the conjunctiva are: 1. Its close adherence to the subjacent parts, being so closely tied down to the tarsi as to possess no wrinkles nor folds; 2. There are wanting in this region of the conjunctiva the glands which are present elsewhere in every mucous membrane.

The second division of the conjunctiva, the cul-de-sac or the palpebral folds, is different from the palpebral portion in being connected with the parts beneath it by a loose connective tissue, by the formation of folds, and by the presence of glands, which will be described hereafter. The upper reflected fold is not easily seen, but by everting the lid, and the eye being directed downward, and the border of the everted lid pressed in the direction of the orbit, a view may be obtained of it. Toward the outer angle of the eye the fold is wanting, but instead there is behind the outer commissure or lid-band somewhat of an excavation. At the inner angle of the eye this reflected portion forms a duplicature of the conjunctiva, called *plica semilunaris*, which contains a minute plate of cartilage (the rudiment of the third lid of animals); its papillæ are small, velvety, and have but little prominence.

It has resting on it a little elevated body, the *caruncula lacrymalis*, which is an aggregation of sebaceous glands, similar to the Meibomian glands, and hair-follicles. These glands are surrounded by fat-cells, and have a size of $\frac{1}{8}'''$ to $\frac{1}{4}'''$. The hairs are very short and quite fine, and have a length of $1'''$ to $6'''$, and $0.'''006$ to $0'''.01$ in thickness. The *plica semilunaris* rests on the inner lid-band (*tendo-oculi*), and supports the *caruncula lacrymalis*. This fact was alluded to in the description of the lid-band.

The third division of the conjunctiva, the *conjunctiva sclerotica*, covers the sclerotica on the lower and inner segment of the globe $3'''$, and on the upper and outer segment from $5\frac{1}{2}'''$ to $6'''$. On the inner and outer sides of the ball it passes $2'''$ back of the insertion of the muscles, and on the upper and lower surfaces about $1\frac{1}{2}'''$. It is more tender and thin than the above described divisions; is somewhat transparent, so that the sclerotica and Tenon's membrane are seen through it. It is rich in elastic fibres, and its submucous connective tissue is abundant, with more or less fat-cells, and is attached to the membrane of Tenon. It is loose and yielding, and permits considerable sliding between the conjunctiva and globe. It lacks papillæ and glands, but has a fully developed epithelium. At the border of the cornea is a ring-formed ridge, $\frac{1}{2}'''$ to $1'''$ broad, which is especially visible in aged persons (*annulus conjunctivæ, limbus conjunctivæ*). It infringes on the cornea more above and below than at the sides, and forms the boundary between the third and the fourth divisions of the conjunctiva.

According to W. Krause it is constituted by the continuation of the fibre-bundles of the conjunctival cellular tissue, which passes on the superior and inferior borders of the cornea. Between these irregularly interlaced bundles, which form slight band-like projections, are longitudinal furrows, which are completely filled by a thick layer of pavement-epithelium. On their transverse diameters, these bands have a papillary aspect, as seen in Fig. 71, and the nerve-fibres end in their cellular tissue by terminal corpuscles. On the contrary, the nerve-fibres destined for the cornea lose their double contour and become very pale on having passed on this membrane. They

FIG. 71.

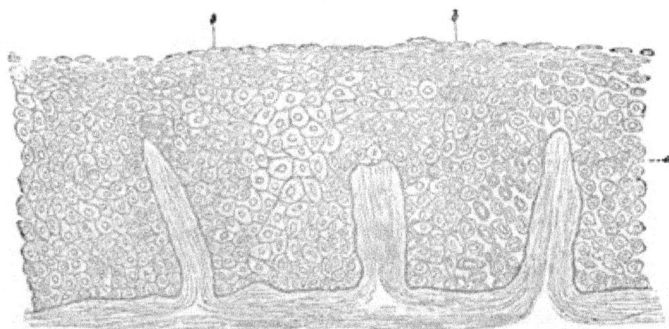

Vertical section of the annulus conjunctivæ of man at the superior surface of the cornea. Magnified 250 diameters. The tract of the cellular tissue of the conjunctiva (discovered by M. Manz), which is prolonged on the corneal border, appears under the form of papillæ (*a*), between which is found a thick layer of stratified pavement-epithelium (*b, b*). (*From Krause.*)

also interlace, but do not form terminal loops. Most likely the isolated fibres always terminate by a club-shaped enlargement, which has no special envelopment. The nerves of the cornea are exclusively beneath the epithelial layer of the anterior corneal surface.

The fourth division of the conjunctiva (*the conjunctiva cornea*) has been described in connection with the cornea.

The description of the arteries and veins of the conjunctiva has been given in connection with the vascular system of the eye, to which part of this treatise the reader is referred.

The nerves are numerous, and are branches of the *palpe-*

bral rami of the *sub-trochlearis*, frontal and lachrymal, which
all proceed from the first branch of the fifth pair. The conjunctiva of the bulb receives its nerves from the sub-trochlear
nerve. The mode in which the nerves terminate in the conjunctiva is peculiar (W. Krause). The branches which emanate
from the sub-conjunctival cellular tissue, by successive division,
their anastomoses, and their exchange of fibres, represent a
rich nervous plexus, of which each ramuscule contains a number of smaller and smaller fibres, whilst the intermediate
meshes become more narrow as the nerves become more
superficial. The nervous fibrils of double contour often divide
dichomatously; they never end in loops, are not lost free in
the tissue, but, on the contrary, they always end by small
particular organs, which Krause has named *terminal clavate
corpuscles (corpuscules clariformes, Endkolben, corpusculæ ner-
vorium terminalia bulboidea)*. These terminal clavate corpuscles
are composed of an envelope of fine cellular tissue, with granular
contents of soft consistence, semi-liquid, and in each of these
corpuscles terminate one or two nerve-fibres of double contour,
which often make several circumvolutions, and represent a large
knot, as seen in Fig. 72. In the interior of these clavate corpuscles the nervous branches divide again into two or three
very fine branches, short and pale, which run a somewhat tortuous course, and terminate in a slight club-shaped enlargement,
as seen in Fig. 73. The clavate corpuscles are always situated
superficially beneath the epithelial layer of the conjunctiva;
their diameter is 0.03 mm. to 0.07 mm., the medium being
0.04 mm. When they are elliptical their length is double that
of their breadth. Krause has named the pale nervous fibres
which are in the interior of these terminal corpuscles, *terminal
fibres*. They have a diameter of 0.0028 mm. The terminal
clavate corpuscles are also found in the lips, the tongue, and
palate in the human being.

In the ocular conjunctiva of man there are found, for each
eye, from 76 to 82 terminal clavate corpuscles; generally one
may be counted to the square millimetre. In the cul-de-sac

and the tarsal portion they are much more thinly distributed. In this last part they are subjacent to the papillæ.

Fig. 72.

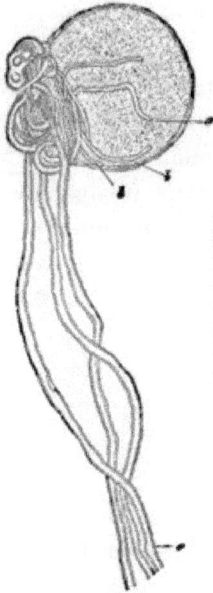

Clavate corpuscles of the conjunctiva of the bulb. Magnified 350 diameters. It is taken from the conjunctiva of man, three hours after death. The nerve-fibres of double contour (c) proceed side by side, and, after having formed a knot, interlace several times before entering into the large corpuscle. a, b, b. Terminal nerve-fibres.

Fig. 73.

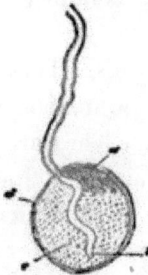

Clavate corpuscles of man eight hours after death, from the ocular conjunctiva. Enlarged 350 times. a. Tortuous terminal fibre. b. Termination of this fibre in form of club. c. Fine granular substance of the corpuscle. d. Nuclei of the cellular envelope. (From Krause.)

The *lymphatic vessels* are quite numerous in the conjunctiva of the bulb, and less numerous in other parts of the conjunctiva. Krause says that in order to discover them under the microscope, it is necessary to make colored injections.

At the corneal border the lymphatic vessels form a delicate

network, with uneven meshes, formed of very fine ramifications, 0.004 mm. in diameter. Where the ramifications anastomose are found enlargements; toward the cornea this network terminates mostly by slightly curved arches. The part of the lymphatic vessels which extends over the breadth of a millimetre is called the lymphatic circle.

At its periphery is found a lymphatic vessel of a larger calibre, and which seems to limit it, and the corneal border is thus surrounded by a somewhat regular circle. To this vascular circle a large number of lymphatics are again connected, which proceed to the centre of the cornea in a radiary direction. These vessels have a diameter of 0.94 mm.; they anastomose among themselves by transverse branches of a diameter of 0.018 mm. to 0.054 mm.

At the distance of 4 to 5 millimetres from the corneal border the vessels which had to that point a radiary direction, take another course; in the upper lid they run parallel to the border of the cornea, and inward and outward, acquire considerable dimensions, and open in the true lymphatic branches, which only are armed with valves. These are directed toward the external and internal angles of the eye, and end in the superficial sub-maxillary lymphatic ganglia. Throughout, the capillary bloodvessels are nearer to the conjunctival surface than are the lymphatic vessels.

The conjunctiva contains *lymphatic glands*. These are follicles, globular or elongated, and completely shut, and are situated immediately beneath the mucous surface. They are composed of an envelope of solid cellular tissue, and of a fine capillary network, which expands in the globular cavity; between the meshes of this network is suspended a second network of cellular tissue, very solid, but finer than the first. The spaces that the two kinds of network leave among them are filled with a little liquid, and a great number of pale cells, round, with only one nucleus, which are perfectly identical with the lymph-corpuscles. The lymph-follicles generally have a diameter of $0'''.2$; in the cul-de-sac they are scattered, and are found, as well in the superior as in the inferior lids, but exist exclusively in their internal half. They are in all respects similar to the solitary glands of the intestinal canal, which have no excretory orifice, and like them, are in connection by their situation with the lymphatic vessels.

Fig. 74.

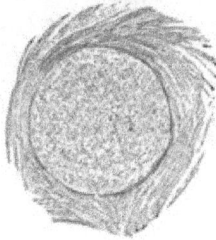

Lymph-follicles, from the third lid of a hog. Magnified 120 diameters. (*From Krause.*)

They are quite variable in number, and sometimes cannot be discovered at all. Bruch first attracted attention to these follicles. Bendz and Stromeyer considered them pathological products, and that they are the seat of trachomatous diseases. W. Krause denies this, and says they are wholly physiological in their character, and that the granulations of military ophthalmy have a quite different structure. W. Krause says that the results of recent injections leave no doubt as to the lymphatic character of these follicles. A fine lymphatic network completely envelops the follicles.

The tissue which surrounds the follicles is, in fact, so rich in lymphatics, that, when examined without having injected the vessels, it seems to be completely filled with lymphatic corpuscles, whilst really these elements are contained within these vascular walls. W. Krause names it lymphatic infiltration; His names it adenoid tissue. There are also numerous lymphatic vessels in the palpebral conjunctiva, according to Krause. Frey has also discovered that between and on the follicles there is expanded a rich network of lymph vessels. It can hardly be doubted that these follicles have a connection with the system of lymph vessels.

The *acinus glands* (*acinus glandulosus*) of the conjunctiva
were first •discovered by C. Krause, and described by him as
glandulæ aggregatæ muciformes. On more thorough investigation
of their character by W. Krause it was ascertained that these
glands are constantly found in the human conjunctiva, and
that they are in their anatomical characteristics entirely similar
to the lachrymal gland. They are found in the cul-de-sac, or
reflected portion of the conjunctiva, between the tarsal cartilages and the eyeball. In the upper cul-de-sac there are 42 of
them, and in the lower cul-de-sac from 2 to 6. They are
located irregularly in the texture of the conjunctiva or beneath
it, and are most numerous in the reflected fold. Near the outer
angle of the eye there are sometimes in the upper cul-de-sac 8
to 12 in a row. In size they are very different (Fig. 75).
They measure from $\frac{1}{3}'''$ to $\frac{1}{5}'''$ to $\frac{1}{15}'''$, and only become visible
under the microscope. Their size seems to be dependent on
their number, the individual glands being smaller the more
numerous they are in the eye.

The form of these glands in the conjunctiva is ordinarily
round or oval. Sometimes two glands are united so as to have
one outlet. Each gland has an oblique outlet, opening on the
conjunctiva. They consist of longitudinal fibres of connective tissue, between which are embedded elongated oval nu-
clei. Their breadth is $\frac{1}{10}'''$ to $\frac{1}{20}'''$, and $\frac{1}{4}'''$ to $\frac{1}{4}'''$ in length.
The excretory duct divides itself into smaller branches. The
acini lie in the expansions of the wall of this canal. Each
acinus is surrounded by a structureless *membrana propria*,
$\frac{1}{700}'''$ in thickness. The acini themselves have a diameter
of $\frac{1}{10}'''$, and can be seen only under the microscope. The
contents of the vesicles consist of cells, free nuclei, and fat-
globules. The cells inclosed within the acini are flat, irregular
polygonal in form, in size $\frac{1}{150}'''$ to $\frac{1}{130}'''$, and contain nuclei of
$\frac{1}{800}$ to $\frac{1}{700}'''$ in size. The neighboring arteries send but few
branches to those glands. These form a large-meshed net-
work. Kleinschmidt could never trace any nervous filaments
to the acini. W. Krause once saw a nervous fibril pass be-

tween two acini, and traced its course for some distance. In another instance he saw a nervous branch enter a large gland, which divided into 8 fibrillæ, which he traced to the middle of the acinus, but failed to discover its ultimate distribution. As far as is now known, the lymphatic glands and the acinus glands compose all the glands of the human conjunctiva. The sweat glands and glands of Manz (see Kleinschmidt, *Archiv*, ix–iii) have not, with any degree of certainty, been discovered in the human conjunctiva.

Fig. 75.

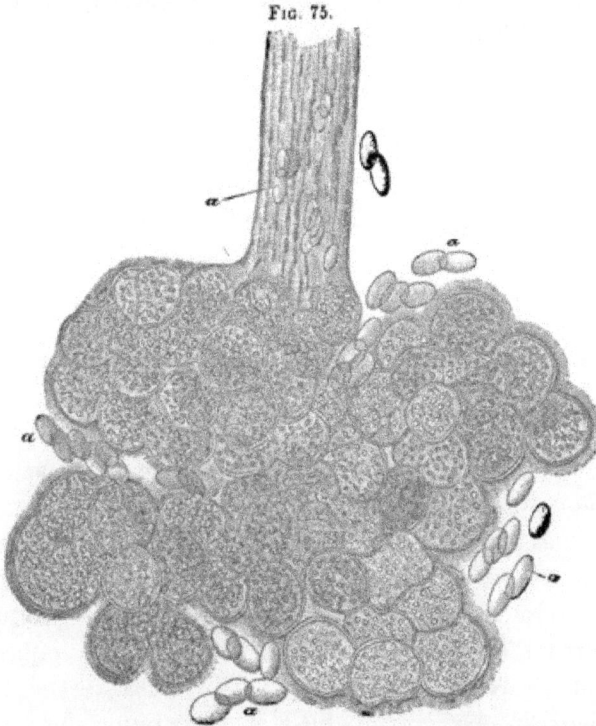

Part of an acinus gland from the conjunctiva of man, magnified 250 diameters. It shows the contents of the acini, and the structure of the outlet of the gland, showing cellular tissue with nuclei. *o, a, a, a, a*. Drops of fat. (*From Kleinschmidt*.)

Glandula Lacrymalis.

The lachrymal gland is lodged in a depression at the outer angle of the orbit, on the inner side of the external angular process of the frontal bone.   It consists of an upper and larger portion, and a lower and smaller portion.   The former (sometimes named *glandula innominata Galeni*) is lodged in the fossa of the zygomatic process of the frontal bone, under the roof of the orbit, has a yellow-red color; in its long diameter it is flat, upward and outward convex, downward and inward concave. Its length is 9''', breadth 5''', and 2½''' in thickness, and weighs 10 to 12 grains, and has a volume of 57 cubic lines.   The second, lower portion (*glandulæ congregatæ Monroi*) lies below the upper portion, and extends beneath the ligamentum palpebrale externum, is 4 to 5''' long, 3½''' broad, 1''' thick, and weighs 3⅔ grains, with a volume of 19 cubic lines.   In structure, they both consist of roundish gland-vesicles, which are somewhat firmly connected by a short cellular tissue, and are enveloped by a common connective tissue membrane.   The individual glandular bodies contain the vesicular beginnings of the smallest excretory ducts, which unite into large branches, and are in number from 8 to 10, and penetrate the conjunctiva in the reflected portion, toward the outer part of the upper eyelid.   Their contents are diffused over the anterior part of the globe.   It is believed by some (Sappey) that the lower lobe sends an excretory duct to communicate with that of the upper lobe.   Hyrtl says that the lower, outer, has several excretory ducts that open into the lower reflexion of the conjunctiva, near the outer eye-angle, so as to supply the lower lid also with the lachrymal secretion.

The excretory ducts, from 8 to 10 in number, consist of structureless membranes, which extend from the conjunctiva to the structureless gland-vesicles.   Their inner surface is lined by cylindrical epithelium, whilst on their outer surface they are surrounded with connective tissue, with some elongated nuclei.   Muscular fibres are wanting, but elastic fibres are numerous, which likely aid in the removal of tears.

The lachrymal gland derives its blood from the *arteria lacrymalis* (Fig. 61), a branch of the *arteria ophthalmica*.   The venous blood is poured into the ophthalmic vein, from the lachrymal vein (*l*, Fig. 61). The nervous supply is wholly from the first branch of the trigeminus, which seems under the microscope to be richly supplied with fine sympathetic fibres. This nerve presides over the secretion of tears, which is wonderfully increased by mechanical irritation, or certain emotional impressions. Under ordinary conditions, there is but little secretion of tears, the greater amount of secretion being the product of the conjunctiva. The lachrymal gland and the conjunctiva are the secretory organs of the lachrymal apparatus. The tears are pure water, with some table-salt and albumen mixed with it, and by analysis, contain, according to M. Frerichs (Krause):

Water, ...	99.06
Solid constituents, ...	0.94
	100.

The solid parts are:

Epithelium, ...	0.14
Albumen, ...	0.08
Chloride of sodium,	
Alkaline phosphates,	
Earthy phosphates,	0.72
Fat and extractive matter,	
	0.94

The acinus glands likely secrete a product altogether similar to that secreted by the lachrymal gland.

The *derivative parts* of the lachrymal apparatus remain to be described. They are the *lachrymal canals*, the *lachrymal sac*, and the *nasal duct*.

The *derivative* lachrymal organs begin on a slight elevation, or papilla, the *papilla lacrymalis*, situated on the posterior edge of the free border of the eyelids, at the outer extremity of the *lacus lacrymalis*. At the apex of each papilla, or tubercle, is a small orifice, the *punctum lacrymale*, which are the commencement of the lachrymal canals (*canaliculi lacrymales*). The papilla are composed of contractile cellular tissue, closely felted, and are not contracted nor dilated, neither spontaneously nor from irritants. The upper punctum has a diameter of $\frac{1}{4}'''$, and is always located a little further inward than the lower, which is a little larger than the upper.

The *lachrymal canals (canaliculi lacrymales)* have a length of $3'''$ to $4'''$, and in diameter they are $\frac{1}{3}'''$ to $\frac{2}{3}'''$. They begin at the puncta, the upper running a short distance vertically upward, and the lower a short distance vertically downward, and with their outer walls are attached to the tarsi, so that they are not only kept in a state of tension, but are always kept open. Their posterior walls are attached to the conjunctiva of the lids. From this point they are enveloped in the connective tissue of the lids, and form an angle, bending convergently inward to open into the lachrymal sac, beneath the lower half of the inner lid-band (*ligamentum canthi internum*), sometimes separately, and sometimes united into one canal. They enter it always quite obliquely, so that their mouths are closed by a fold of mucous membrane. The lower is $1'''$ shorter than the upper, and larger in diameter.

Their inner surface is lined by a mucous membrane, which is tender, pale, smooth, with few mucous glands, and is covered by a lamellated pavement epithelium. These canals are freely surrounded by the fibres of the lachrymal portion of the orbicularis, in the manner described when treating of that muscle. As rare exceptions, two puncta have been observed in one lid.

The *lachrymal duct (ductus lacrymalis)* lies behind the frontal process of the superior maxillary bone, in the *canalis lacrymalis*, and opens into the inferior meatus of the nose. In consequence of its course backward, it forms, with the floor of the nasal cavity, an angle of $65°$, and to the vertical meridian an angle of $5°$ to $10°$. It is about $1'''$ in length, and is divided into 3 parts. (1) The lachrymal sac (*saccus lacrymalis*); (2), the part surrounded by a bony canal,—the *pars maxillaris ductus lacrymalis;* and (3), into the lower, nasal portion, the *canalis naso-lacrymalis.*

The *lachrymal sac* is $5'''$ to $6'''$ in height, and $2'''$ to $3'''$ in breadth, of an elongated, oval, or almond form, being flattened from before backward to such an extent that often, in the cadaver, the two walls are closely pressed together (Stellwag). One-half lies in the *fossa lacrymalis*. More than one-half of its vertical expansion is below the rim of the inner, lower border of the orbit. The upper half of the sac is, for some distance, crossed by the lid-band. The upper cul-de-sac, or fundus, passes 1½''' above the upper border of the *ligamentum canthi internum*. Behind this ligament the lachrymal canals perforate its outer wall. According to Arlt (*Compte rendu du Congrés*, 1863), only the upper third of the sac is covered by the fibres of the lachrymal portion of the orbicularis (Horner's muscle). The inner wall of the portion of the sac within the lachrymal fossa, passes down vertically, and without any change, into the inner wall of the membranous portion of the lachrymal duct. Arlt says, that in many instances, the outer wall of the sac, before opening into the bony portion of the duct, forms a sinus or recessus. In cases where this sinus does not exist, then there is also no mark of division on the outer wall. In some instances there is a marked projection of the mucous membrane at the point of division between the sac and nasal duct, so that there is a marked constriction at this point. In such instances there is greater development of the periosteum, or of the aponeurosis of the sac, at the point of entrance into the bony canal.

The *maxillary portion* of the lachrymal duct is surrounded by a bony canal, and is connected in its whole extent by the periosteum. It is most constricted in the middle in a length of 3''' to 4''', and has a diameter of ¾''' to ⅝''' Like the lachrymal sac, it always seems to be filled with fluid.

The nasal portion.—The lachrymal duct does not terminate with the bony canal at the inferior meatus of the nose, but runs along 2''' to 3''' between the bony wall and the mucous membrane of the nose, and with an oblique, narrow opening it perforates the nasal mucous membrane at an acute angle. The nasal portion then is only covered by a fold of the mucous

membrane of the nose. Its nasal mouth is 7''' to 9''' from
the anterior border of the frontal process of the superior maxillary, and is 3''' to 5''' above the floor of the nose. The
mouth into the nose is longer from above down than transversely, and varies from a slit ¾''' long to ⅔''' wide, to an oval
opening of 2''' long to 1''' to 1½''' wide. It is always filled
with fluids.

It often happens that at the opening into the nasal passage,
small duplicatures of the mucous membrane exist, which proceed either from above, from before and behind, or from behind,
but always lie flat on the Schneiderian membrane. These
cause the slit-formed mouth to be either horizontal, oblique,
vertical, or bent in the form of a bow.

The *ductus lacrymalis* has a thick mucous membrane, which
is rough, and is lined by a simple *epithelium ciliare;* in the lower
part, however, it possesses a lamellated pavement *epithelium,*
and has numerous *racemose mucous glands.*

In its whole extent, the duct is surrounded by a network of
firm connective tissue, which possesses elastic fibres. This
network is unusually rich in bloodvessels. These vessels,
which are connected with the surrounding bone, also richly
supplied with blood, fill up the interspaces, and in the cadaver
they do not collapse, so that they can be seen with the naked
eye. This stratum of vessels is thin at the lachrymal sac, but
becomes much thicker along the nasal duct, especially posteriorly, so that the duct is narrowed, and the mucous membrane is projected inward and thrown into *folds.*

Outward the connective tissue becomes firm and tendinous,
which envelops the duct as a sheath in its whole extent. As
far as the bony canal extends this sheath is loosely connected
with it, and performs the function of periosteum.

On the outer wall of the lachrymal sac it forms a kind of
aponeurosis, which is attached to the edge of the lachrymal
fossa, which makes of it a closed canal or cavity. This aponeurosis is in intimate connection with the processes of the
posterior surface of the lid-band, as well as with the sheath of

the lachrymal portion of the orbicularis, through which it is strengthened.

The bloodvessels and nerves of the lachrymal apparatus are mostly branches supplying neighboring organs. The lachrymal gland has a branch from the arteria ophthalmica, the *arteria lacrymalis;* also a corresponding vein, which empties into the vena ophthalmica. The nerves are derived from the nervus infra-trochlearis, a branch of the first ramus of the nervus trigemini, which, when irritated mechanically from without, or from within *emotionally*, causes the rapid secretion of tears.

INDEX.